The Institution of
Engineering and Technology

Code of Practice

Electrical Energy Storage Systems

2nd Edition

Publication information

Published by The Institution of Engineering and Technology, London, United Kingdom
The Institution of Engineering and Technology is registered as a Charity in England & Wales
(no. 211014) and Scotland (no. SC038698).

The Institution of Engineering and Technology,
Michael Faraday House,
Six Hills Way, Stevenage,
SG1 2AY, United Kingdom.

Copies of this publication may be obtained from:
The Institution of Engineering and Technology
PO Box 96, Stevenage, SG1 2SD, UK
Tel: +44 (0)1438 767328
Email: sales@theiet.org
www.electrical.theiet.org/books

While the publisher and contributors believe that the information and guidance given in this work is correct, all parties must rely upon their own skill and judgement when making use of it. Neither the publisher nor contributors assume any liability to anyone for any loss or damage caused by any error or omission in the work, whether such error or omission is the result of negligence or any other cause. Any and all such liability is disclaimed.

The moral rights of the authors to be identified as author of this work have been asserted by them in accordance with the Copyright, Designs and Patents Act 1988.

A list of organizations represented on this committee can be obtained on request to IET Standards. This publication does not purport to include all the necessary provisions of a contract. Users are responsible for its correct application. Compliance with the contents of this document cannot confer immunity from legal obligations.

It is the constant aim of the IET to improve the quality of our products and services. We should be grateful if anyone finding an inaccuracy or ambiguity while using this document would inform the IET Standards development team at ietstandardsenquiries@theiet.org or the IET, Six Hills Way, Stevenage SG1 2AY, UK.

ISBN 978-1-83953-041-8 (paperback)
ISBN 978-1-83953-389-1 (electronic)

Typeset in the UK by the Institution of Engineering and Technology, Stevenage
Printed in the UK by Sterling Press

Contents

Contents

Contents

Lists of Figures and Tables

Lists of Figures and Tables

List of Figures

Lists of Figures and Tables

Lists of Figures and Tables

List of Tables

Acknowledgements

The IET gratefully acknowledges the advice and assistance provided by the following people and organizations in the development of this Code of Practice.

Technical Authors:

Martin Cotterell BEng(Hons) FIET
EUR ING Graham Kenyon BEng(Hons) CEng MIET (G Kenyon Technology Ltd)

Special thanks to:

Dr Andrew Crossland (Advance Further Energy) for additional example schematics (Figures G.2, G.4 and G.5)

Steven Fletcher (Frazer-Nash Consultancy) and Professor Dani Strickland (Loughborough University) for assistance in the approach to DC arc flash risk assessment (Appendix E).

Additional Contributors:

Arash Amiri (Loughborough University)
Kiran Arora (Bryan Cave Leighton Paisner LLP)
Sue Bloomfield (RECC)
Jonathan Buston (HSE Science Division)
Aikaterini Chatzivasileiadi (Cardiff University)
Frank Gordon (REA)
Nicky Herbert (BEIS)
Peter James (Lyra Electronics)
Geoff Kingston (BEAMA)
David Lindsay (Certi-fi)
Ian McDonald (Connected Energy)
Nik Mitchell (Certsure)
Tim Oldershaw (J Brand)
Luke Osborne (ECA)
Emmanouil Perisoglou (Cardiff University)
Chris Roberts (MCS)
Steve Saunders (Arup)
Rob Such (RS Renewable/Wattstor)
Richard Tullett (Sonnen Batterie)
Cliff Wilson (Advance Further Energy)

 **Section 1**

Introduction

Electrical energy storage systems (EESS) provide storage of electrical energy so that it can be used later. EESS may be installed for a number of reasons, generally for one or a combination of the following main applications.

(a) Increasing the 'self-consumption' of buildings fitted with renewable energy systems

For example, for buildings fitted with renewable generation systems, such as wind, solar PV or hydro-electric, there will be times during the day, varying with the time of year and with weather conditions, when the power generated by the renewable generation is greater than the power requirements of the building. During those periods, all the surplus electricity – in a system without storage – is exported into the grid (or curtailed by the generation inverter in a zero export scenario).For systems with storage, some of the energy that would otherwise be exported can be retained for use later in the day – an operation sometimes called 'time-shifting'.

(b) Arbitrage services

By charging a battery during times when electricity costs are lower, and providing power at times of peak electricity prices or on demand, such as during distribution use of system 'red zones' (the highest tariff periods), an EESS facilitates the reduction in annual electricity costs for commercial and industrial customers. Arbitrage is also possible for domestic customers on certain tariffs, where use of the EESS is managed by brokerage services.

(c) Ancillary services

EESS are able to provide grid support features such as assisting with frequency regulation or voltage support. They can also be used to manage export constraints when there are high levels of distributed generation deployment and to defer costly electricity network investment or reinforcement.

(d) Providing a back-up or alternative power supply

An EESS can provide electricity in the event of a power cut and can also be used to provide a dedicated supply for specific loads.

Note: This is just a summary of the main applications – a more comprehensive description of EESS applications is covered later in Section 4.

This Code of Practice looks at EESS applications and provides information for practitioners to safely and effectively specify, design, install, commission, operate and maintain a system.

The second edition takes account of developments in the industry, and progress in standardization.

Section 2

Scope, purpose and use of this Code of Practice

2.1 Scope and purpose

The purpose of this Code of Practice is to provide:

(a) a reference to practitioners on the safe, effective and competent application of EESS;
(b) an understanding of the common terms and operating modes of EESS; and
(c) detailed information on the specification, design, installation, commissioning, operation and maintenance of an EESS.

The scope of this Code of Practice includes EESS intended for fixed installation applications, including:

(a) individual dwellings;
(b) commercial applications, including multi-occupancy buildings and multi-occupancy residential buildings; and
(c) industrial applications.

It covers:

(a) electrochemical energy storage systems in electrical installations;
(b) integration into low voltage (LV) power systems (AC and DC); and
(c) systems aligned with existing standards, regulations and guidance.

The following items are out of the scope of this Code of Practice:

(a) application of EESS into high voltage (HV) power systems (for example, distribution or transmission networks);
(b) product standards;
(c) supply chain issues;
(d) consumer level guidance;
(e) thermal energy storage (either where the storage is used to support a thermal system or where the storage is reversible and provides electricity); and
(f) systems utilizing batteries contained within an electric vehicle (an EV; for example, 'vehicle to grid' systems). It is recognized that those implementing vehicle to grid systems will need to consult this Code of Practice along with IET *Code of Practice for Electric Vehicle Charging Equipment Installation*, Section 10, which discusses vehicle to grid systems.

Note: While this Code of Practice is focused on LV EESS, some parts would apply to the design and installation of HV systems.

Section 2 – Scope, purpose and use of this Code of Practice

2.2   Use of this Code of Practice

This code of practice is arranged as follows:

	Informative introduction to EESS:
Sections 1 to 6	Section 1: Introduction
	Section 2: Scope, purpose and use of this Code of Practice
	Section 3: EESS components and architectures
	Section 4: EESS operating states and applications
	Section 5: Batteries
	Section 6: Other EESS components

	Normative requirements for those using the Code of Practice to design, install, operate and maintain EESS:
Sections 7 to 14 **N** Normative	Section 7: EESS safety and planning considerations
	Section 8: Specification of an EESS
	Section 9: Design of an EESS
	Section 10: Network connection and DNO approval
	Section 11: EESS installation
	Section 12: EESS inspection and testing
	Section 13: EESS handover and documentation
	Section 14: EESS operation and maintenance

	Informative supporting information:
Appendices	Appendix A: Glossary
	Appendix B: References
	Appendix C: System labels and safety signs
	Appendix D: Standards
	Appendix E: DC arc flash risk assessment
	Appendix F: Frequently asked questions
	Appendix G: General example schematics

 # Section 3

EESS components and architectures

3.1 Overview

This section aims to provide an outline of a typical EESS and its configurations to set the scene for topics discussed in other sections of this Code of Practice.

3.2 System functionality

The functionality of a particular EESS will depend on a number of factors including:

(a) the purpose of the system;
(b) the system architecture;
(c) whether the EESS is intended to provide DC or a combined (DC and AC) power supply;
(d) whether the EESS is designed to link to a local generator (for example, a solar PV system);
(e) whether the EESS is a pre-manufactured packaged system or a custom-designed system assembled from relevant components for a particular application;
(f) whether the system is intended for operation independently of the grid;
(g) whether loads are to be segregated into critical and non-critical, so that high power or non-critical loads can be load-shed in island mode operation; and
(h) whether the EESS is to provide uninterrupted power supply with defined performance and power continuity parameters.

3.3 Power coupling modes

Recognizing the way in which the EESS is to be connected to electrical energy sources and loads is fundamental to its design and specification.

With AC coupling, the input or output operates at the grid supply voltage and frequency, typically 230 V AC 50 Hz for single phase LV installations and 400 V AC 50 Hz for three-phase LV installations. Since the energy storage is a DC source, this necessitates one or more items of power conversion equipment (PCE) and a battery management system (BMS) to charge the storage battery from an AC source. PCE is also needed to power AC loads and export to the grid, synchronized with the grid supply where necessary.

Note: In systems where the EESS is exclusively used as an addition to a renewable generation scheme, it is common for DC coupling to refer to a battery being connected on to the DC circuits of the renewable generation system (whether or not directly or via PCE). AC coupling refers to the use of an EESS with separate PCE, and connection of these to the AC power systems separately.

Section 3 – EESS components and architectures

When using an EESS with a renewable system (such as solar PV), the main differences/considerations between AC and DC coupled designs are illustrated in Table 3.1.

Table 3.1   Differences between AC and DC coupled design

	AC coupled	DC coupled *
Equipment	Wider range of compatible components more readily available, as most LV installations operate from standardized AC voltages (to BS EN 50160).	Where the system is LV DC, cable losses may mean that the renewable system's PCE must be electrically compatible with the EESS components to be fitted, and also with DC coupling in the EESS. The sizing of the existing PCE needs to be checked; it is usually the case that the renewable system's grid-connected inverter is sized to suit the renewable generation.
	EESS is wholly separate to a grid-connected renewable system. The system can be arranged so that is it not wholly dependent on one PCE; if one PCE fails then either the generation or the storage will continue to operate.	EESS is an integral part of the grid-connected renewable system.
	Inverters in AC systems can increase harmonics. There are harmonic limits set in product standards for inverters.	Fewer problems with switching harmonics in DC systems.
	AC coupled system has an output with standard AC voltages (to BS EN 50160) and may be easier to integrate into an existing system.	The existing renewable inverter may need replacing. This may not be an issue if the existing inverter is at the end of its service life.
	Locating the battery away from the renewable system and inverter is generally simpler. Provides more choice in battery location.	Battery may need to be located adjacent to renewable system components, particularly for ELV batteries. Less choice in battery location or requires longer DC cabling or the use of higher DC voltages.
	EESS and renewable energy system inverters can each be sized for their own system requirements (no need for oversizing either of these).	Where peaks are diverted into the EESS, a smaller inverter can be used by the renewable system if desired (cost savings), although this may introduce complexity and require expensive intelligent controls. With existing UK export tariffs, the renewable energy would ideally be exported at the maximum export rate, when the battery is full, to prevent curtailment.
Circuits	Residual current devices (RCDs) are readily available, and newer types are unaffected by DC residual currents. AC systems, circuits, protection and earthing are familiar to many designers and those who will work on systems.	RCDs are not readily available, although standards are being developed. Some DC systems operate at lower voltages, perhaps at extra-low voltage (ELV). To deliver the same power, larger cross-sectional area conductors and interconnection components are required. See Section 9.10. Suitably rated switching and protection is required for DC circuits, due to the difficulties of extinguishing DC arcs. See Section 9.10 DC systems, circuits, protection and earthing may not be as familiar, as DC systems are currently not as widely used.

Table 3.1 *Cont.*

	AC coupled	DC coupled *
PV – battery charge efficiency	Increased losses: more power conversion stages (electricity from renewable generator gets converted first to AC then back to DC to charge battery).	Reduced losses: less conversion stages between renewable generator and battery. May require DC to DC converters.
Feed-in income	No effect on feed-in income: Can use a standard generation meter arrangement (meter located on inverter output).	Reduced feed-in income: as no DC kWh generation meters are currently accredited, an AC meter is used on the output of the whole system. Hence, for energy that passes through the battery, the losses in the charge/discharge cycle will reduce the feed-in income.
* These comments refer to typical DC coupled systems, using an inverter designed to fulfil this function. It is also possible to have a DC coupled system using a standard renewable grid-connect inverter and a DC to DC converter. This may add additional losses to the EESS, but does allow the use of a standard inverter and opens the door for retrofitting.		

3.4 Packaged EESS and discrete component EESS

3.4.1 Overview

A brief comparison of the benefits and drawbacks of different types of EESS is shown in Table 3.2.

Table 3.2 Benefits and drawbacks of different types of EESS

Type	Potential benefits	Potential drawbacks
Packaged EESS (Section 3.4.2)	'One-stop-shop' solution. Manufacturer takes responsibility for extensive type-testing and ensures that the product can deliver its stated performance. Manufacturer includes and tests all relevant safety functions, such as thermal and electrical safety protection, within the package. Operators and maintainers can follow manufacturer's instructions.	Interface options, bespoke options and the range of system performance may be limited. Tied to 'manufacturer-recommended' or supplied replacement components.
Discrete component EESS (Section 3.4.3)	Wider choice of interface options. Easier to make the installation bespoke to needs. Freedom to choose products from different manufacturers.	Greater reliance on the designer and installer of the system to correctly specify, source, install, test and commission. The design and commissioning stages would require more documentation than other types. The designer and installer need to compile operation and maintenance information to ensure that the 'as-delivered' system can be safely operated and maintained.
UPS, CPS, or simple DC battery backup system (Section 3.4.4)	Can be distributed throughout the electrical installation, closer to the point(s) of use or on selected circuits, to provide power in the event of a fault within other parts of the electrical installation, as well as loss of the grid supply.	Costly for smaller-scale commercial users and in dwellings. Generally independent of locally generated energy. Typically not suitable for grid-demand support or profile-shifting.

3.4.2 Packaged EESS

A packaged EESS is a complete solution available commercially as an off-the-shelf product. The packaged system need not comprise a single enclosure. For example, a packaged system may comprise separate energy storage units with separate PCE.

It is the designer's responsibility to ensure that a packaged solution meets the requirements of all relevant legislation. The designer is recommended to follow the guidance given in this Code of Practice.

The system may be purpose-built for use with a specified rating or ratings of local generation system (solar PV or wind).

Importantly, the controller and battery are specified by the manufacturer of the system. This specification is based on a pre-defined set of input and output power conditions. Interface options, and the range of system performance, may therefore be limited. Consequently, it is important to ensure that for the most part the load, generation, and grid supply profiles of the installation are suited to the packaged system as appropriate.

A key advantage of a system of this type is that the manufacturer's installation, operation and maintenance instructions are readily available to ensure safety during use, maintenance and decommissioning. Depending on the manufacturer's lifecycle policies, components for repair may be more readily available than with bespoke systems.

3.4.3 Discrete-component (assembled on site) EESS

In assembling an EESS from discrete components, the system designer has a choice of interface options, and is not limited to features and performance of a single manufacturer.

However, there is a greater reliance on the competence and experience of the designer to ensure the system matches the installation load, grid and generation profiles, and is correctly specified, sourced, tested and commissioned. The designer also has greater responsibility for the safety of the system as a whole in all of its operating modes, and during maintenance and decommissioning. Additionally, relevant documentation and schematics are required to be compiled into a suitable operating and maintenance manual, including relevant user instructions.

Maintainers may encounter issues with obsolescence of certain components of the EESS, since product lifecycles of components may not be aligned, even for products from the same manufacturer. It is therefore imperative to include design and selection parameters within the operating and maintenance manual to ensure that compatible replacement components can be safely selected.

3.4.4 UPS, CPS and simple DC battery backup systems

Uninterruptible power system (UPS), central power supply (CPS) and simple DC battery backup systems often serve certain areas, equipment or circuits in the installation, rather than the installation as a whole, and are often most effective when they are distributed throughout the electrical installation.

In addition to their basic function of providing power to certain critical or sensitive items of equipment during power outages and fluctuations, they often have the added means of conditioning power.

They are, however, generally independent of locally generated power and, due to the limitations of their typical battery technology in charge/discharge cycles, may not be suitable for grid-demand support or profile-shifting.

Section 3 – EESS components and architectures

3.5 Examples of system architectures

3.5.1 Grid-connected system without any other local generation sources

Perhaps the simplest type of grid-connected EESS, there are two broad categories:

(a) grid-connected EESS, which can charge the battery from the grid, for example, when energy is cheap. This type of EESS provides energy to the loads, and/or export back to the grid, at peak usage times. Such architecture is illustrated in Figure 3.1.

(b) UPS, CPS or DC battery-backup system. This application is discussed in Section 3.4.4. Figure 3.2 shows examples of a typical UPS system, and Figure 3.3 shows a typical DC battery backup application. Arrangements for CPS are provided in BS EN 50171.

Figure 3.1 Example of grid-connected systems without other local generation

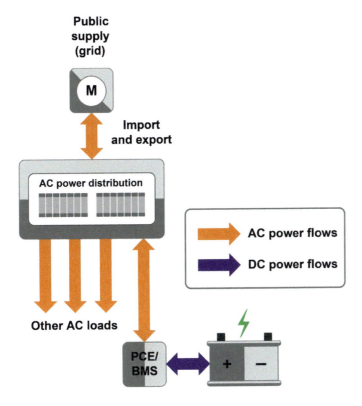

Figure 3.2   Example UPS applications without other local generation

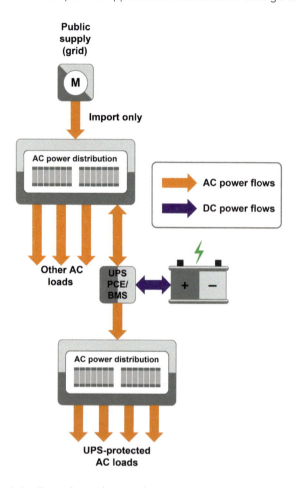

Figure 3.3   Example DC battery backup applications without local generation

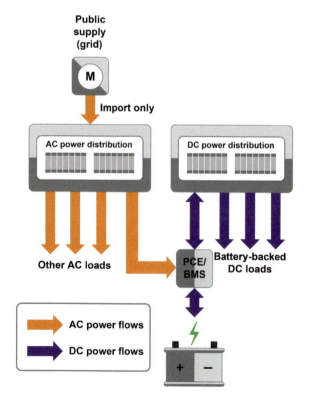

Section 3 – EESS components and architectures

3.5.2 Grid-connected system with other local generation

This type of system may provide the most flexibility for the consumer.

The choice of architecture will depend on a number of factors, including whether:

(a) the other local generation is existing, or is being installed at the same time as the EESS;

(b) the EESS is being specified to charge solely from the other local generation and/or charge from the grid; and

(c) the other local generation and EESS will be AC or DC coupled.

The issues raised above have a number of technical architecture solutions. Examples are provided as follows:

(a) Figure 3.4 shows a grid-connected system that is AC coupled, and both the EESS and other local generation feed into a distribution board.

Note: In practice, they do not necessarily need to feed into the same distribution board or consumer unit.

(b) Figure 3.5 shows a grid-connected system, in which the storage (charge/discharge and battery) parts of the EESS are DC coupled to other local generation PCE.

(c) Figure 3.6 represents a similar system to that in Figure 3.5, which permits charging of the storage from the grid in addition to other local generation.

Note: Where a battery is coupled to a metered generation system, it needs to be ensured that energy taken from the grid, stored in the battery and later returned to the grid or installation is not recorded on the total generation meters, by ensuring meters and charging circuits are connected at appropriate points. Failure to do this could be considered fraudulent and may void feed-in (or equivalent) payments. Further information can be found in guidance on storage and feed-in tariffs from the Office of Gas and Electricity Markets (Ofgem): https://www.ofgem.gov.uk/system/files/docs/2018/12/storage_guidance_final_v2.pdf

(d) Figures 3.7, 3.8, and 3.9 show examples of how UPS and DC battery backup systems may be used in installations with other local generation.

Figure 3.4   Grid-connected system, other local generation, AC coupling

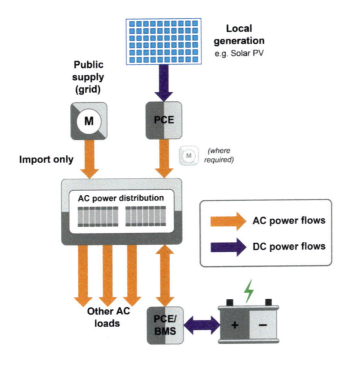

Section 3 – EESS components and architectures

Figure 3.5 Grid-connected system, other local generation, DC coupling

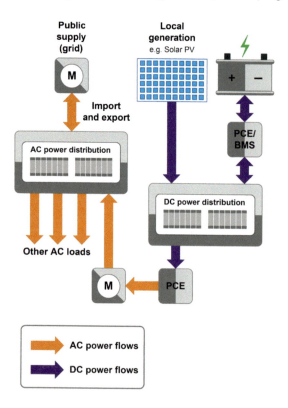

Note: In this arrangement, the PCE and the BMS may be integrated within a single device.

Figure 3.6 Grid-connected system, DC coupling and mains charging

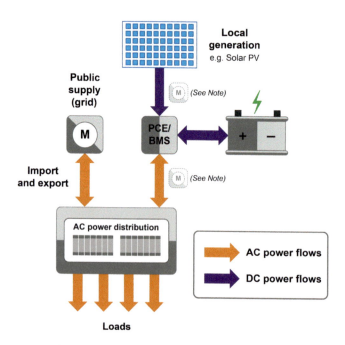

Note: Where a battery is coupled to a metered generation system (for example, where an export tariff arrangement is in place), the installer must ensure that the battery that is charging from the mains is not recorded on the total generation meters. Failure to do this could be considered fraudulent.

Section 3 – EESS components and architectures

Figure 3.7 Example UPS application with other local generation

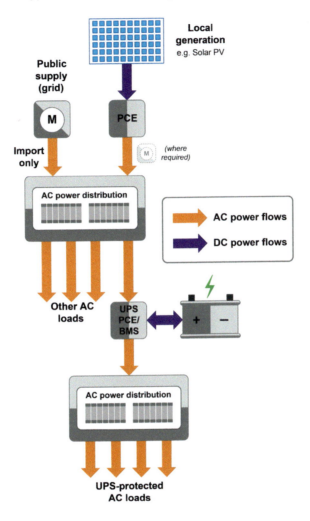

Section 3 – EESS components and architectures

Figure 3.8 Example UPS application with a DC coupled EESS and other local generation

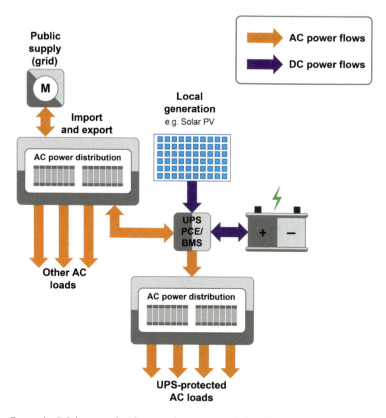

Figure 3.9 Example DC battery backup applications with local generation

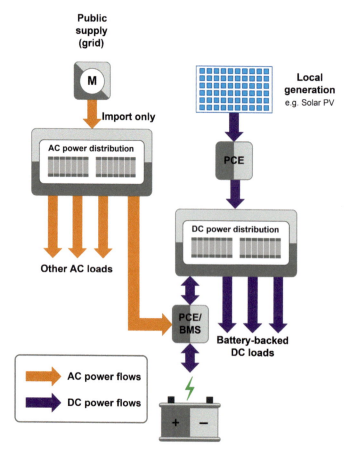

3.5.3 Grid-independent systems

Grid-independent systems operate independently of a network operator's supply. A local generation source, for example solar PV, wind or rotary generator, is often used as a primary power source. Where loads are very low power, it may be possible to have a grid-independent system that uses only secondary batteries charged at another location, or through a mobile means of charging. Grid-independent systems may be suitable for:

(a) using renewable generation (such as solar PV) only, with DC loads (see Figure 3.10). Examples of this application include:

    (i) lighting of road signage that uses energy efficient lighting; and

    (ii) remote monitoring or telemetry units, or security applications.

(b) using renewable generation, possibly supported by other generation such as a backup generator (see Figure 3.11). Examples of this application include:

    (i) mobile or transportable units, such as road maintenance and construction welfare units;

    (ii) telecommunications equipment assemblies in remote locations; and

    (iii) mission or safety critical remote monitoring or telemetry units.

Figure 3.10   Grid-independent system with DC loads

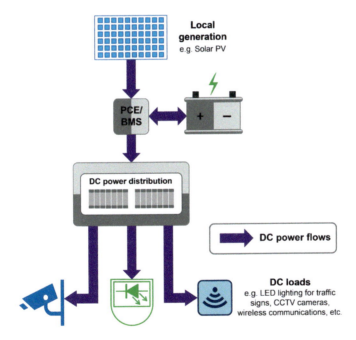

Section 3 – EESS components and architectures

Figure 3.11   Grid-independent system with AC and DC loads

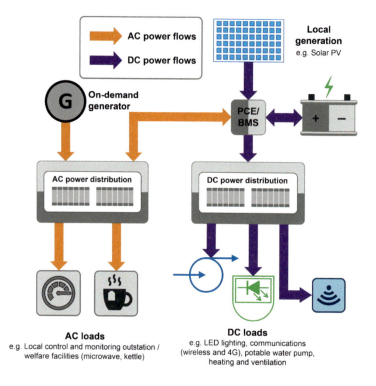

AC loads
e.g. Local control and monitoring outstation /
welfare facilities (microwave, kettle)

DC loads
e.g. LED lighting, communications
(wireless and 4G), potable water pump,
heating and ventilation

 **Section 4**

EESS operating states and applications

4.1 Introduction

This section describes some common applications of EESS.

4.2 Optimizing self-consumption of renewable energy

4.2.1 General

Where buildings contain renewable systems, such as wind or solar PV, there will be times of the day and/ or year when the power generated by a renewable system is greater than the power requirements of the building. For buildings without an EESS, the surplus energy generated is exported to the grid. Where the building has an EESS, some or all of the energy that would otherwise be exported or curtailed can be retained for use at a later time.

4.2.2 Example of a basic time-shifting operation with solar PV

One of the applications for an EESS is in addition to a solar PV system – storing electricity that would otherwise have been exported to grid. The electricity stored in the battery can then be used at a later time. This process is often termed 'time-shifting', but 'energy shifting' or 'increased self-consumption' are also used.

Section 4 – EESS operating states and applications

Figure 4.1 Example of basic time shifting operation © IET 2017 (Reproduced as adapted from *Battery energy storage systems with Grid-connected solar photovoltaics – A Technical Guide (BR 514)*; © IHS Markit 2017)

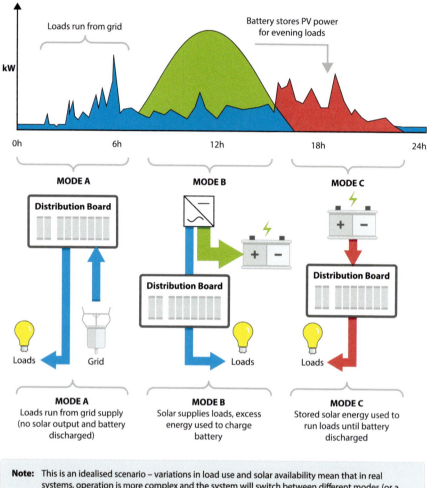

Note: This is an idealised scenario – variations in load use and solar availability mean that in real systems, operation is more complex and the system will switch between different modes (or a combination) on multiple occasions during the day.

Hence in an idealized situation, a solar-storage system operating in time-shifting mode can be thought of as:

(a) start of the day (early morning): running from the grid;
(b) as the sun comes out: shifting to a stage where the solar energy is used to power the loads and recharge the battery; and
(c) in the evening: using the power stored in the battery to run the loads.

Once the system has determined that the battery is fully discharged, the system switches back into grid-only mode, ready to start the whole cycle again.

During the UK winter months, the average sunlight hours are much lower than in the summer months, and the system may therefore have more than one 'programme' for how it controls:

(a) the use of renewable energy;
(b) the use of energy from the grid; and
(c) the charge/discharge cycle of the battery.

This is required to optimize use, and may also be necessary to prevent damage to the battery by deep discharge or over-cycling.

4.3 Arbitrage and aggregation

Arbitrage describes the process of using the grid supply to charge storage batteries when electricity tariffs are lower, such as at daily low demand periods, thus reducing the costs of electricity at peak demand periods when tariffs are higher.

Where local generation is also available, the EESS controller must be optimized to ensure that the arbitrage operation maximizes the use of available local generation. It may mean, for example:

(a) the daily arbitrage charge contribution varies seasonally; or
(b) where solar PV is used and the arbitrage charging is overnight, the arbitrage charge contribution varies both seasonally and in response to the charge state of the storage battery at dusk.

Aggregation describes the remote management of several storage systems, usually totalling around one megawatt and often located in different geographical areas, by a third party under contract with the National Grid to supply both excess electricity at peak times and excess capacity at off-peak times.

Installers should be mindful of the consumer's requirements, and potential future requirements, when installing EESS, so that income and savings potential over and above self-consumption of locally generated electricity is considered.

For installations in commercial or industrial premises, the arbitrage settings and parameters may be programmed based on the needs of the installation (for example, switched off or reduced during shutdown periods or holidays).

4.4 Reserve network capacity

Infrastructure elements can and do fail. There can also be unforeseen temporary increases in demand.

Many networks, both public and private, have reserve generation, but this may take time to run up from the failure occurring. EESS in a supply network can be used to provide a temporary reserve of power to cover these periods. As well as adding to security of supply, they may also be used to defer the need for a supply upgrade (see also Section 4.7).

4.5 Island mode, grid-independent, UPS and CPS operation

4.5.1 Introduction

The grid supply may be interrupted or operate outside of its usual parameters. A supply complying with BS EN 50160 or the Electricity Safety, Quality and Continuity Regulations (ESQCR) 2002 is permitted to deviate from the specified ranges of voltage and frequency for a short period of time, but this may not suit all loads. The EESS can be configured or, in the case of UPS, be dedicated, to providing a specific quality and continuity of power to specific loads when the grid supply is interrupted or operated outside optimal parameters.

4.5.2 Island mode operation

In this state, grid power is disconnected completely from all, or part of, the installation, which is then supplied by local generation and/or stored energy. The EESS can also automatically disconnect from the external supply when quality of supply is outside limits thus protecting loads from poor power quality.

Section 4 – EESS operating states and applications

Where the total load in the installation exceeds the capability of the inverter and/or storage, it is necessary to 'load shed' high power loads, and provide power to critical loads during the period the system is operating in island mode.

4.5.3 Grid-independent systems

Grid-independent systems are those where the installation is designed to be permanently isolated from the grid. In such systems, the available local generation will be specified to provide sufficient power to charge the storage batteries sufficiently so that they do not run low. Where loads are running at the same time as charging is underway, the local generation must also provide this power.

4.5.4 UPS or CPS

The aim of a UPS is to provide continuity of power to the load. Often, the UPS is tasked with providing conditioned power to specific critical loads, such that the load is not exposed to disturbances on the normal supply, and continues to operate when the normal supply is interrupted. The inverter of the UPS will provide power to the load most of the time, only switching to the normal supply during faults or maintenance.

Ideally, the UPS will keep batteries charged until they are needed by the load and hence the batteries used in UPS have different characteristics from those used in other types of EESS.

A CPS used to provide continuity of power for specific emergency or safety system applications such as emergency lighting.

4.6 Ancillary services

Note 1: Whilst this section discusses voltage support and frequency response, it is necessary to maintain both system voltage and AC frequency within certain limits in public supplies. It is also necessary to maintain utilization voltages and frequencies for equipment connected to private supplies, within limits considered by the equipment manufacturer.

Note 2: Some of these services can be supported by vehicle to grid systems. The electrical safety requirements for vehicle to grid systems are outside the scope of this document, but are covered in the IET *Code of Practice for Electric Vehicle Charging Equipment Installation*.

4.6.1 Voltage support

For grid supply or heavy-use private networks, the ability to manage system voltage and maintain it within certain limits is an extremely important consideration. For public networks, the objective is to comply with regulatory limits, and in private networks, to optimize system efficiency.

EESS can be used to achieve both these objectives by either importing or exporting real and reactive power.

4.6.2 Frequency response and the balancing market

As demands on a supply network vary, the frequency may increase or decrease accordingly. In a similar manner to voltage support, regulatory or efficiency optimization can be achieved by EESS embedded in the network so that during over-generation real power can be imported. Likewise, real power can be exported when the demand on the network increases.

The current minimum size of generation for participation in the balancing market has been reduced to 1 MW, and is now feasible for aggregated EESS participation.

4.7   Network upgrade deferral and peak lopping (peak shaving)

Due to the demand at peak times, distribution network infrastructure (for example, distribution cabling and transformers) may approach or exceed their design limits. Provision of EESS close to points of peak demand usage may alleviate the strain on the infrastructure at peak periods and hence delay or remove the need for upgrading the infrastructure. The EESS is charged at off-peak times, and discharges at peak times, thus reducing the load currents transferred across the infrastructure.

Where demand at peak times in a consumer's installation exceeds agreed supply capacity for a short period, the use of an EESS can help the consumer avoid financial penalties. Such operations are not new, and consumers have employed peak-lopping schemes using local generating sets. The use of an EESS for this purpose helps avoid the storage and use of fossil fuels (oil and gas) on site.

Where the natural evolution of the consumer's site installation means the demand at peak periods may increase, it may be possible to use EESS in a similar manner to Network Upgrade Deferral (NUD), thus avoiding a costly supply upgrade. This would need to be factored in to the design of a peak-lopping system.

4.8   Renewable integration and export limitation

Renewable integration permits the installation of more renewable generation (or indeed higher peak loads) where the grid is constrained. This may concern, for example, locations where voltage or reverse power limits are causing localized grid issues, or micro-grids where there is insufficient demand to integrate renewable generation at peak output.

For systems where the power available from renewables for export may exceed the export limits plus the local demand, energy can be stored in an EESS for export later when the generated power is lower, for example, to improve self-consumption or to permit more generation in areas with grid constraint or to allow price arbitrage.

 **Section 5**

Batteries

5.1   Common storage devices

There are many different battery types, both primary (non-rechargeable) and secondary (rechargeable). Some of these have been around many years while others are at the research and development phase. A small subset of the latter is likely to come onto the market in the future. Each of these battery types has different characteristics. To help determine the trade-off between cost and performance of different battery chemistries, it is important to consider these characteristics in conjunction with the application. The characteristics are usually obtained from the manufacturer. However, caution is strongly recommended when it comes to claimed manufacturer battery performance statements. Where possible, independent evaluation should be obtained to verify battery performance.

Table 5.1 summarizes key characteristics that may be required for an application and gives examples of common battery types available today that claim good performance in those areas. It should be noted that these are claimed and not validated. Some battery types in use today, such as nickel-cadmium, are toxic; as such, their future use is likely to be limited, and they are therefore not listed in the table.

As battery chemistry is continually evolving, it may be worth investigating the suitability and track record of new chemistries for applications at the time of design if considered appropriate.

Table 5.1   Typical battery characteristic considerations

Example characteristic		Application	Example battery type	Notes
Life span	Cycle life (frequent deep discharge)	Arbitrage, daily peak load lopping or export limiting.	Some forms of lithium-ion (e.g. lithium-titanate), vanadium flow battery.	Claimed high cycle life at deep discharge (suggested 20-year life with daily operation).
	Cycle life (occasional deep discharge)	Back-up systems or UPS.	Lead-acid.	Although the cycle life is not high, if the battery will not be undertaking frequency cycling, then this cheaper option may well be more cost effective. Full recycling is available for lead-acid batteries.
Charge/ Discharge	Rate (fast rate of charge/ discharge)	Ancillary services such as enhanced frequency response, island mode or grid-independent applications.	Nickel-metal-hydride, lithium-ion.	Batteries that are able to go from charge to discharge at a fast rate will be more useful for these applications. In some applications a supercapacitor is also an option.
	Self-discharge	Battery back-up or NUD.	Flow battery, lead-acid.	Any battery that is required to sit for long periods will self-discharge. This may result in the need for continuous trickle charge.
Other	Energy density	Where size or space may be an issue.	Lithium based batteries such as lithium-ion or lithium-polymer.	This measure of performance is more common in the automotive industry.

Other characteristics to be aware of when deciding on battery type include, but are not limited to:

(a) voltage;

(b) maximum current;

(c) calendar life (the life of the battery regardless of whether or not it is in use);

(d) overcharge tolerance, which can damage the cells (nickel-metal-hydride and lithium-ion batteries have low tolerance);

(e) operating temperature (for example, nickel-metal-hydride may be better in cold environments);

(f) maintenance and the need for recalibration;

(g) safety and additional protection (for example, temperature monitoring); and

(h) impedance of the battery (which should be used to check protection).

There is a significant amount of literature available on the internet that offers non-validated comparisons of characteristics between different battery types. In this ever-changing landscape, as different batteries appear on the market, it is the responsibility of the designer to look carefully at all information available in order to make an informed decision.

5.2   Battery characteristics

In order to design an EESS, it is necessary to understand the characteristics of the energy storage device, together with its operational and environmental requirements. As well as influencing overall system design and sizing, these characteristics can significantly impact on the selection of other system components.

It should be noted that some batteries have particular temperature characteristics for charging and discharging, and may need additional equipment for battery management, ventilation and cooling. This additional equipment must be installed to ensure that the battery warranty conditions are met.

5.2.1   Nominal capacity

Information on nominal capacity is provided by the manufacturer and describes how much energy the battery can nominally deliver from fully charged, under a certain set of conditions. The main factors that influence capacity are:

(a) the battery type and size;

(b) the charge/discharge rate;

(c) the operating temperature;

(d) the state of health of the battery (which is influenced by usage and maintenance); and

(e) the number of charge/discharge cycles it can deliver.

Note: Manufacturers' assumptions on aging projections may differ.

Traditionally, battery capacity was described as an Ah (ampere-hour) figure at a particular discharge current. However, in many EESS applications, there has been a shift to using kilowatt hour (kWh) or megawatt hour (MWh) as the means of quoting capacity.

Manufacturers can use significantly different operating assumptions when quoting battery performance data. Depending on the application, some parameters of manufacturer performance data will be more relevant than others. For example, where a discharge/charge cycle of once per day or more is required, lifetime capacity, and consequential cost per kWh, will be more relevant than in a purely 'back-up power' scenario.

When comparing batteries, it is important to obtain the assumed operating parameters/conditions relevant to the stated capacity figure to make sure you are comparing like with like.

5.2.2   Hour rating

Batteries may have an hour rating, which describes both the maximum rate at which the battery can be charged and the length of time for which the battery can be discharged at the stated full discharge rate.

For example, a 1-hour battery can be discharged at its stated discharge rate for 1 hour, whereas a 2-hour battery can be discharged at its stated discharge rate for 2 hours.

Section 5 – Batteries

5.2.3 C-rate

There are two types of C rating in common usage, Cx and xC, which are explained in this section.

5.2.3.1 Cx ratings

To enable a simple comparison between batteries, the discharge current is expressed in terms of a C-rate. The C-rate normalizes the discharge current relative to the battery capacity. For example:

(a) a rate of C1 equates to the current that will completely discharge the battery in 1 hour. For a battery rated at 500 Ah at the C1 rate, this relates to a discharge current of 500 A.

(b) a rate of C20 equates to the current that will completely discharge the battery in 20 hours. For a battery rated at 200 Ah at the C20 rate, this relates to a discharge current of 10 A (200÷20).

Note: C-rate can also be used in a similar manner to describe charge rates.

In general, the quicker a battery is discharged at the same current flow, the smaller its capacity. A real example from a commercially available deep cycle lead-acid battery is shown in Table 5.2.

Table 5.2 Battery capacity variations for a 468 Ah, C20 battery

Rate	Capacity (Ah)	Discharge current (A)
C100	605	6.05
C20	468	23.4
C10	398	39.8

The capacity of most deep cycle lead-acid batteries are described at the C20 rate. The C20 or C10 rate is generally a good figure to use when considering batteries for a grid-independent solar PV system.

Determining the C-rate for the quoted capacity is very relevant when comparing batteries. For example, as can be seen in Table 5.2, one supplier may describe a battery as 398 Ah, while another supplier may describe the same battery as 468 Ah – this sounds like an 18 % 'bigger' battery, but is in fact the same unit.

Similarly the C-rate used in sizing calculations must be relevant to the application. For example, there is no point in using C100 data if the battery is expected to be fully discharged over a couple of hours.

5.2.3.2 xC ratings

Alternatively, an xC rating can be used (the figure precedes the letter C). The xC rating indicates how quickly the EESS can charge or discharge. For some battery types, the xC rating when charging may be different to the xC rating when discharging.

A battery rated at xC is expected to be able to fully charge/discharge in 1/x hours. For example, a battery rated at 2C is expected to be able to deliver its full rated capacity over a period of ½ hour.

While usage varies, the xC rating can also be used to express the relationship between the energy and power ratings of an EESS. For example, an EESS formed from a 100 kW inverter coupled with a 200 kWh battery that is able to fully charge/discharge over a period of 2 hours is denoted as ½C (0.5C or C/2).

5.2.4 Depth of discharge

Depth of discharge (DOD) describes how fully a battery has been discharged during a discharge cycle. It is expressed as a percentage of nominal battery capacity, for example, 60 %. A discharge of around 80 % represents 'deep cycle' operation.

DOD can have a significant impact on battery life span, particularly for some technologies such as traditional lead-acid. In general, the higher the DOD, the shorter the life span. Hence for many systems, it is relatively common to program the system so as to limit the routine DOD.

Deep cycle operation is required for many EESS. Some battery chemistries (for example, lead-acid) may be available in both shallow and deep cycle variants. Other types may be suitable for deep cycle operation (for example, most lithium-ion batteries can be discharged to around 80 % of nominal capacity without significant effect on battery life span).

Note: Lead-acid batteries are the oldest type of rechargeable battery. However, while the chemistry may be broadly the same, the design of a lead-acid battery varies considerably depending on the intended application. This includes modifications to the shape, size and composition of the electrodes and to the formation of the electrolyte. For example, traditional automotive batteries tend to be constructed from many thin plate electrodes (to maximize the surface area for chemical reactions and to deliver high cranking currents – so as to start the car); whereas deep cycle batteries designed for solar applications have larger, thicker and more robust electrodes (for a longer life and a deeper DOD).

Automotive starter batteries are not generally suitable and are not advised for use in solar PV storage applications.

5.2.5 Effective capacity

As it is common for a system to be programmed to limit the DOD, the term 'effective capacity' (sometimes termed 'usable capacity') is often used to describe the usable capacity of the battery. The effective capacity is usually less than the nominal (name-plate) capacity.

For example, a battery with a 500 Ah capacity, on a system programmed to limit DOD to 60 %, the effective capacity is 500 × 0.6 = 300 Ah.

Note: This effectively means that, as a customer, you have to purchase a battery with a greater kWh rating than you will routinely use.

Some manufacturers are choosing to only quote the effective or usable capacity of the battery, particularly where the use case and charge/discharge rate is well understood and well controlled by the EESS. As an example, a manufacturer may quote a battery as having "7 kWh to 100 % DOD". This is designed to make selecting a battery, and understanding its use, easier for the customer.

5.2.6 Battery round-trip efficiency

All batteries are subject to some losses during the charge → storage → discharge cycle. The round-trip efficiency describes how effective a battery is throughout the full cycle.

For example, for a battery with a 90 % round-trip efficiency, for every 100 kWh put into the battery, only 90 kWh will be usefully returned, the residual 10 kWh being absorbed by losses (usually heat).

The efficiency is influenced by a number of factors including:

(a) battery type;
(b) battery temperature;
(c) the rate of charge and discharge;
(d) the storage interval; and
(e) the age of the battery.

5.2.7 EESS round-trip efficiency

While battery charge–discharge efficiency is an important factor, it is the overall efficiency of the complete EESS (battery, battery controls, inverter, charger, thermal management systems etc.) that is key.

Section 5 – Batteries

When determining what EESS round-trip efficiency figure to use (when, for example, modelling systems), it is necessary to ensure that the efficiency figure applies to appropriate system operational parameters. Factors such as ambient temperature may not only impact on the battery efficiency, they may also lead to more energy being lost in the EESS thermal management systems (cooling or heating the battery and ancillary components).

5.2.8 Fault current

Fault current is the current that a battery can deliver into a short circuit across its output terminals. Battery fault currents can be significant and it is important to ensure that circuit protective devices are able to safely and effectively interrupt this current. Overcurrent protective devices should be selected and co-ordinated with batteries in accordance with BS 7671. Chapter 3 of IET Guidance Note 6 *Protection Against Overcurrent* provides guidance on protection against fault current and the breaking capacity of overcurrent protective devices.

Note: Overcurrent protective devices, such as circuit breakers or fuses, have both a current value they will operate at (which varies with time – as expressed on the time-current graph for a particular protective device) and also a maximum interrupt rating (the maximum fault current the device can interrupt). If the battery fault current is greater than the interrupt rating of the protective device, the device may not be able to interrupt the fault current effectively. The DC maximum interrupt rating of an overcurrent protective device is often different to the AC maximum interrupt rating.

Dependent on the arrangement of batteries, DC arc flash may result from batteries with significant fault current. Where this is the case, and the risk has not been assessed by a product manufacturer of a packaged product, an arc flash risk assessment is recommended. See Appendix E.

Note: Arc fault detection devices (AFDDs) to BS EN 62606, as referenced in BS 7671, are for single-phase AC final circuits only. They are not specifically aimed at preventing arc flash events, and would not prevent DC arc flash.

5.2.9 Maximum power

Many batteries have a maximum continuous current they can deliver. This may be due to a limit within the battery chemistry or from the battery control/power handling circuits.

5.2.10 Self-discharge rate

Self-discharge describes a normal characteristic of all batteries to gradually lose charge over time. The degree of self-discharge varies with battery type, age and temperature. Battery safety and control circuits also contribute to the standing losses.

5.2.11 Battery life

See Section 5.4 on battery warranties.

5.3   Battery charge profiles

Batteries all have an ideal charge profile that ensures they reach a fully charged state quickly and efficiently. This charge profile varies between battery chemistries.

In many EESS, the nature of the application means that it may not be possible for the system to follow the ideal charge profile (for example, a solar EESS may – on many days – not have sufficient solar resource to perform a full charge cycle). However, knowing the ideal profile for a particular battery type can significantly help when designing a system and understanding EESS performance. In addition, the battery charge profile may need to be considered where ancillary services are to be delivered against a pre-defined contract.

Section 5 – Batteries

A battery is made up of a collection of appropriately connected smaller batteries and cells. To optimize battery life, some battery types require cell charge management and/or cell balancing for each cell. In some cases, the way in which a battery bank is electrically arranged can help to balance the current/voltage of cells, such as making power connections at diametrically opposite terminals (using exactly the same length of wire in each string, and ensuring consistent connections and terminations) as illustrated in Figure 5.1.

Figure 5.1 Example power connections to a battery bank

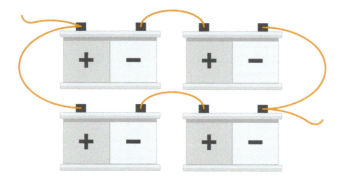

Some types of battery are more prone to cell reversal, where a cell in a battery having a lower capacity than the other cells, due to a manufacturing defect or degradation in use, is driven into a reverse condition during rapid discharge. The risk of cell reversal in such battery types can be addressed by:

(a) arranging batteries in parallel strings;
(b) using lower DC string voltages; and
(c) monitoring voltage in battery strings and stopping discharge charge if reversal is detected.

5.3.1 Three-stage charge profile

The following diagram (Figure 5.2) shows a typical three-stage battery charge cycle. This three-stage cycle is generally the preferred method for charging and maintaining the life span of lead-acid batteries.

Figure 5.2 Typical three-stage battery charge cycle

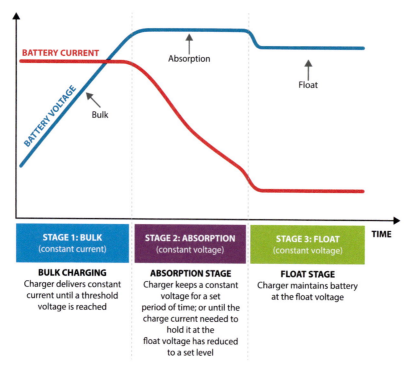

Section 5 – Batteries

Some batteries also require a periodic equalization charge. This is a controlled over-charge, where the charger holds the battery at a high voltage (high voltage, low current). Certain battery types require periodic reverse voltage to be applied to limit dendritic growth, which can short out the electrodes. Vented lead-acid (VLA) batteries require a periodic equalization to reduce sulphation of the electrodes and to de-stratify the electrolyte.

Note 1: A dendrite is a growth of metallic crystals, with branches that resemble trees, that can occur on electrodes in some battery types.

Note 2: VLA batteries' fluid levels need to be checked after equalization, as equalization can liberate significant amounts of hydrogen and oxygen.

Note 3: Stratification is where the electrolyte is more highly concentrated at the bottom of the cell than the top. In VLA batteries, stratification occurs when the battery has been kept at charge levels below around 80 %.

Note 4: Equalization is a process of slightly increasing the charge voltage for a short time. It is a technique used with VLA batteries to remove sulphate crystals that accumulate on the electrode plates.

5.3.2 Charge profile for lithium-ion batteries

Precise control of charge voltage is used with lithium-ion batteries. While similar to charging a lead-acid battery, Figure 5.3 shows a typical two-stage lithium-ion charge cycle:

Figure 5.3 Typical two-stage lithium-ion charge cycle; courtesy
http://batteryuniversity.com/learn/article/charging_lithium_ion_batteries

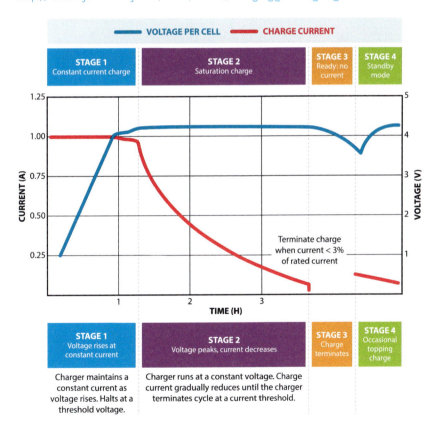

5.3.3 Opportunity charging

'Opportunity charging' describes a charging regime that occurs whenever power is available, such as from the intermittent output of a solar PV system.

The nature of an intermittent charge current means that opportunity charging does not require a battery to be fully discharged before commencing charging; nor does it always continue until the battery is fully recharged.

The effectiveness of opportunity charging can vary enormously. Long-term, repeated incomplete recharge cycles can have a significantly detrimental impact on the battery life span of some battery types (such as lead-acid). Hence, for any EESS, it is important to ensure that the likely charge profile is well understood and matched to a suitable battery type.

Note: Lithium-ion batteries are more tolerant to intermittent partial charging patterns: they respond well to partial charging and do not need a frequent full charge in the same way that most lead-acid batteries do.

To maximize battery life and system efficiency, the BMS built into the EESS is designed to provide a charge profile as close as possible to the ideal for the particular battery technology it is connected to. However, variations in available charge current will have a significant impact on the availability of sufficient charge current to deliver the ideal charge profile.

Figure 5.4 Example of charge current available from an EESS linked to a solar PV system © IET 2017 (Reproduced as adapted from *Battery energy storage systems with Grid-connected solar photovoltaics – A Technical Guide (BR 514)*; © IHS Markit 2017)

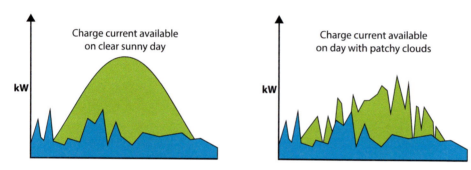

5.4 Battery warranties

5.4.1 General

A battery warranty sets out how long it is guaranteed to function correctly and what battery capacity can be expected over certain time or usage intervals. Ensuring the best practical conditions of operation (environment, charge/discharge, maintenance) helps to preserve battery capacity and thus battery life.

Battery capacity degrades as a result of both time and the nature of how it has been used. Factors that influence the gradual loss in capacity vary with battery type and may include:

(a) calendar aging, dependent on:
  (i) environmental conditions (especially temperature);
  (ii) resting state of charge; and
  (iii) battery age (total number of years in service); and

(b) cycle aging, dependent on:
  (i) DOD (both average and maximum usage);
  (ii) the pattern of charging/discharging in use; and
  (iii) the number of charge/discharge cycles delivered.

Due to the number of factors that can influence capacity degradation, battery warranties can be complex to define and to operate. For example, many battery warranties are defined in terms of both time (number of years in service) and number of full cycles delivered. While 'years in service' is easy to record, determining how many 'full cycles' a battery has delivered can be difficult. Additionally, determining what constitutes a full cycle, and the cumulative effect of multiple part cycles (for example, is two half cycles the same as one full cycle?) can further complicate matters.

Section 5 – Batteries

Capacity warranties are provided in various formats and vary significantly depending on technologies. Designers and end users should take care to compare DOD cycles and life time degradation including taking into account the total energy (kWh) available from the battery during its lifetime when performing within normal operating conditions.

Capacity warranties may give a single figure (for example, "guaranteed to provide 60 % of original capacity after 10 years or 2,000 cycles, whichever comes sooner") or may be stepped – see Figure 5.5.

5.4.2 Information provided to consumers

Installers should highlight the warranty of the battery in terms of both years and cycles to consumers, and explain how both relate to the design of the system that is being installed. This is particularly important for systems in which the projected number of cycles leads to a shorter warranty period than the warrantied life in years.

Figure 5.5   Example of battery capacity warranties

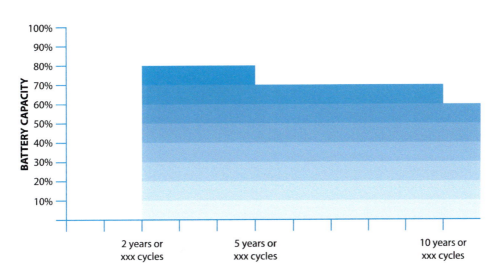

5.5   Hybrid battery systems

Where some performance parameters of a system cannot be met by a specific battery chemistry or architecture that is suited to the system performance most of the time, it is possible to design a hybrid system. For example, some of the performance required by a more expensive technology that responds to large demand changes, but perhaps has a more expensive lifecycle cost and lower energy density, forms part of the energy storage system, and the bulk high energy storage with higher energy density is used for the remainder.

Note: Such systems need not rely solely on battery storage; however, the design of systems with other types of energy storage is outside the scope of this Code of Practice.

Where a hybrid system is planned, it may be necessary to separately accommodate different battery chemistries or architectures, as they may not be compatible for co-location. For example, they may have different:

(a) safe working practices for maintenance;
(b) thermal environment;
(c) fresh air ventilation throughput; or
(d) requirements for fire suppression and fire fighting.

 # Section 6

Other EESS components

This section provides information on the main components, other than the storage device itself, that form an EESS.

Note: The following list of components is not intended to be exhaustive and does not cover general electrical items such as transformers, circuit protective devices, switchgear etc.

While the following devices are described separately, some or all of the functions described may be incorporated into a single device.

6.1 Inverter

An inverter is a type of power conversion equipment (PCE) that converts DC to AC.

6.1.1 Inverter types

Various types of inverter may be found in an EESS:

(a) Grid-connect inverter

A grid-connect inverter sits between a DC source (for example, a battery or PV array) and the grid. It converts the DC source into an AC output that is synchronized with the grid.

As well as synchronizing its output voltage and frequency with the grid, the inverter also monitors the grid and temporarily switches off (disconnects) if the voltage or frequency go outside allowable limits or there is a loss of grid power.

A pure grid-connect inverter will not produce any output without the mains connected on its AC side. Hence, it will not provide any output during a power cut.

Some grid-connect inverters also feature control circuits that can curtail output under certain conditions, for example, where there is site export limit.

Figure 6.1 Grid-connected inverter © IET 2020 (Reproduced as adapted from *Battery energy storage systems with Grid-connected solar photovoltaics – A Technical Guide (BR 514);* © IHS Markit 2017)

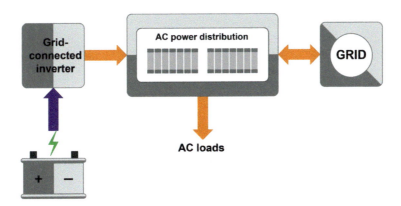

Section 6 – Other EESS components

(b) Stand-alone inverter

A stand-alone inverter is designed to be connected to a battery so that it provides AC power without the grid present.

AC loads can be run directly from a stand-alone inverter but:

    (i) the size/number of loads that can be run is limited by the rating of the inverter; and
    (ii) the length of time the loads can be run for is dictated by the size of the battery.

Figure 6.2 Stand-alone inverter © IET 2020 (Reproduced as adapted from *Battery energy storage systems with Grid-connected solar photovoltaics – A Technical Guide (BR 514)*; © IHS Markit 2017)

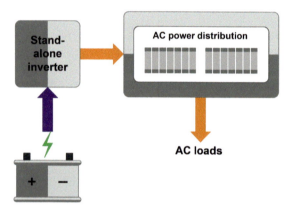

(c) Grid-connect and stand-alone inverter

Inverters are available that can operate in both grid-connect or stand-alone mode. These inverters are used for grid-connected systems that need the additional functionality of being able to operate in stand-alone mode during a power cut.

Additional provisions are required for this kind of inverter – for example, circuits to detect the presence of a grid supply and to ensure that out-of-synchronization re-connection to the grid does not occur.

(d) Bi-directional inverter

A grid-connected inverter that can both draw power from the grid and feed power back to the grid is termed a 'bi-directional inverter'.

Figure 6.3 Bi-directional inverter

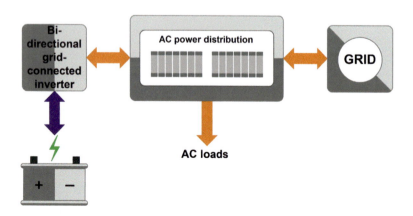

Section 6 – Other EESS components

6.1.2 Inverter characteristics

The following characteristics (see Table 6.1) are useful to consider when selecting inverters:

Table 6.1 Useful characteristics for selecting an inverter

General characteristics	
Minimum and maximum DC input voltage	Inverter must be able to operate at the lowest nominal DC voltage and withstand the maximum DC voltage of the EESS. **Note:** This may need to include a consideration of any voltage variations due to temperature.
Maximum input current	Key parameter when specifying DC circuit components (cables, protective devices, isolators etc.).
Nominal AC output voltage & frequency	For grid-connect inverters in the UK, the minimum and maximum output voltage and frequency that the inverter will operate will be set by G98/G99.
Minimum and maximum grid voltage & frequency	
Electrical separation	Knowing whether the PCE has a transformer that performs galvanic separation can be a key consideration in the design process (e.g. may influence RCD selection; see Section 9.12).
Efficiency	Important considerations when selecting an inverter for any application and ensuring the inverter is suited to its loads (e.g. power factor for inductive loads).The ENA maintains an on-line register of G98/G99 type-tested devices, at: https://www.ena-eng.org/gen-ttr/ **Note:** the ENA on-line register only addresses the G98 and G99 type tests, and does not include type tests for parameters such as efficiency, ingress protection rating, and environmental tests.
Power factor range	
Continuous output rating	
Grid synchronization	
IP rating	
Ambient temperature range	
Max. humidity	
Max. altitude	
G98/G99 type-test status	
Warranty	An important consideration for selecting an inverter (and other components of the EESS).

Key stand-alone inverter characteristics	
Continuous output power	The power output the inverter can support continuously.
Short-term peak current capacity	The ability of the inverter to handle inrush currents of connected equipment (both the magnitude of the inrush, and its duration, need considering).

Key bi-directional inverter characteristics	
Maximum charge current	The maximum charge the inverter can supply.
Charge profile(s)	The types of charge profile (for different battery types) that the inverter is able to provide.

6.1.3 Harmonics

Inverters can generate harmonic content at their output, and input, and this may cause compatibility issues with equipment connected to input and output circuits.

There are broadly three different kinds of inverter, each having a different harmonic profile, illustrated in Figure 6.4:

(a) sine wave, sometime referred to as 'true sine wave' or 'pure sine wave';
(b) modified sine wave; and
(c) square wave.

Section 6 – Other EESS components

Some electronic equipment may be damaged by these harmonics, and these are not suitable for supply from an inverter output, whether operating in parallel with, or as a switched alternative to, the grid supply. For example, some appliances containing electronic controls or variable speed drives are incompatible with connection to supplies that have battery storage or solar PV inverters. Some single-phase induction motors, for example, those in fridges and freezers, may also experience overheating as a result of too high a harmonic content in the supply, for which they have not been designed.

Installers should be aware that this may affect the warranty of certain domestic appliances. They should also make consumers aware of this, so that they can make informed choices when purchasing appliances, and particularly when selecting which appliances can remain connected in island mode.

Some issues with harmonics can be resolved with local voltage optimization fitted at the appliance, but this increases costs for the consumer.

Figure 6.4 Examples of inverter output voltage waveforms

(a) sine wave

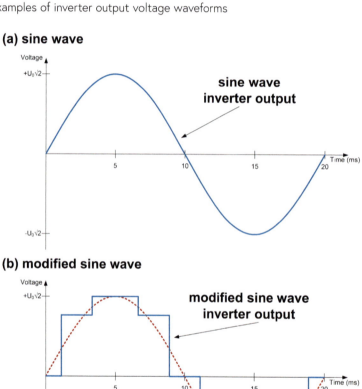

(b) modified sine wave

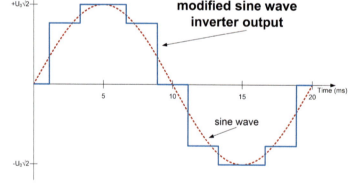

(c) square wave

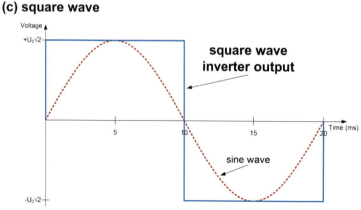

Section 6 – Other EESS components

6.2 Battery management system (BMS)

6.2.1 General

A BMS provides and controls the electricity flowing to the battery during the re-charging phase. The charger manages the whole charge sequence and provides a charge profile appropriate to the battery size and type. This involves precise control of output voltage and current. For many chargers, a temperature sensor connected to the battery provides feedback to the charger to enable it to adjust the charge profile to battery cell temperature.

6.2.2 Charging control and active battery management

Using a charging system that is not suitable for the particular battery type and chemistry can cause damage to the battery and may present a significant safety hazard. The charger and charge control system used in an EESS must be matched to that battery type. In general, there are two types of charger:

1. voltage-controlled chargers, which vary the voltage delivered to the cell or battery; and
2. current-controlled chargers, which vary the current delivered to the cell or battery.

Depending on battery chemistry/construction, the BMS may include cell balancing to optimize the battery and charging performance.

It is essential that lithium-ion cells are actively measured and monitored by a BMS, especially to detect over charging and over temperature, and cease charging before safety parameters are exceeded. Equivalent controls may be required for other battery chemistries or physical arrangements that may be developed in the future.

6.2.3 PCE for battery charging

Three main types of charging PCE control can be found in an EESS:

(a) a DC BMS, which is used as a direct interface between a DC source (such as a solar PV array) and the battery. It converts the variable DC input voltage to the precise DC charge voltage required. It may include PCE to convert DC voltages from one voltage level to another.

(b) an AC BMS, which is designed to provide a controlled DC charge sequence from an AC source – such as from the mains (grid) or a generator – and as such has integral PCE.

(c) a combined PCE/BMS: some inverters combine a number of functions, for example, taking DC power out of the battery when working in 'inverter mode' and operating as a battery charger when operating in 'charger mode'.

Table 6.2 Useful characteristics for selecting PCE for battery charging

Key charge controller characteristics	
Maximum charge current	The maximum current the charger can supply.
Charge profile(s)	The types of charge profile (for different battery types) that the charger is able to provide.
Efficiency	Important considerations when selecting a charger for any application.
Power factor range	
IP rating	
Ambient temperature range	
Max. humidity	
Max. altitude	

6.2.4 Discharge control

As well as needing control of the charging phase, control over discharge is also required. Discharge control functions typically include:

(a) ensuring that discharge is halted at a set battery DOD;
(b) limiting the rate of discharge (discharge current); and
(c) providing temperature feedback (adjusting discharge profile to battery cell temperature).

For grid-connected battery storage systems, the discharge control system may also control and restrict when discharge can occur. Examples include:

(a) restricting discharge to certain time windows during the day; and
(b) preventing discharge until certain recharge/battery voltage thresholds have been reached.

Note: In determining whether to halt discharge, control systems may factor in both the battery voltage and the discharge current (higher currents will cause the battery voltage to fall lower).

6.3 DC to DC converting PCE

Some battery systems use a DC to DC convertor as an interface between the battery and all other parts of the EESS. While this type of PCE adds cost and can reduce efficiency, it is useful in providing:

(a) a means to step up the battery voltage (higher voltage can enable better integration with some inverters and can reduce cable size);
(b) a fixed system voltage (DC to DC output remains constant, while battery voltage varies);
(c) impedance matching between the generation source and storage load to maximize energy transfer;
(d) galvanic isolation of the battery and a ready means of isolating the battery; and
(e) optimization and matching of multiple battery packs within a single EESS.

6.4 Thermal management system

All EESS devices have a temperature band within which they can operate. Both minimum and maximum temperature limits apply.

Most EESS feature some form of thermal control. This can vary from a basic system that simply switches off the EESS if it goes outside the acceptable thermal operating band to more sophisticated systems that can cool (and in some cases heat) the storage device to enable operation.

Note: Guidance on assessing the requirements for fire detection and prevention, and ventilation, can be found in Sections 7and 11.

6.5 EESS controller

The EESS controller monitors the status of the EESS and provides charge/discharge signals depending on the selected control algorithm. This control may be based on an algorithm that runs independently within the controller, but may also be subject to external control signals (for example, from a distribution network operator, distribution system operator, transmission system operator, or aggregator) or from measured changes in external parameters (such as a loss of mains or a change in grid frequency, or tariff information received from a smart meter via the home area network).

The controller may also provide other functions, such as data logging and providing an interface to the web. The EESS may also control external devices such as changeover or isolation relays.

Section 6 – Other EESS components

The EESS controller can be a stand-alone device that communicates with other system components over a data bus. Alternatively the EESS controller may be incorporated into other system devices, such as the inverter.

In high power and/or large storage capacity EESS, a suitable hierarchy of control should be established resulting from the operational and functional risk assessment. Appropriate safety functions should be distributed where the hierarchy of control prefers this. In general, local protective provisions (for example, within individual enclosures, particularly those containing batteries) should be put in place.

The BS EN 61508 series of standards may be a useful tool for the functional safety of controls, monitoring and interlocks.

6.6   Independent means of earthing

In accordance with BS 7671, the EESS cannot rely on the distributor's earth connection during island mode operation.

Where the system will operate in island mode, a means of earthing in compliance with the relevant requirements in BS 7671 and BS 7430 is required.

Grid supplies with TN-C-S or TN-S earthing arrangements are less likely to have an independent local means of earthing within the installation (prior to the installation of an EESS). Consumer earth electrode systems are sometimes used in installations with these arrangements to help limit touch voltages in the event of a broken combined neutral and earth conductor in a protective multiple earthing distribution network. Where there is no suitable consumer earth electrode system available in the installation, an additional consumer earth electrode is required where the system is intended to operate in island mode.

Where the grid supply is TT, the installation may have a consumer earth electrode system in accordance with BS 7671 and BS 7430; however, its suitability for use with the anticipated operating modes of the EESS shall be assessed.

An unearthed EESS may be possible for certain specific cases, for example, a grid-independent system operating as separated extra-low voltage (SELV). However, for general LV electrical installations, it is unlikely that an unearthed system for island mode or grid-independent operation would be suitable, due to:

(a) the prevalence of connected Class I appliances, and electronic devices with functional earth connections, connected to telecommunication networks (internal and external) providing fortuitous earth connections or requiring a controlled discharge path for static that accumulates in copper data network cabling;
(b) the effects of voltage disturbances of atmospheric origin or due to switching in the HV network that may couple with the installation via incoming metallic telecommunication services and require surge protection devices (SPDs) connected to earth; and
(c) the expense involved in installing IT electrical systems, should that method of protection against electric shock be used. Note that IT systems may be suitable for some applications, particularly in industrial settings, or where required for medical locations and similar applications.

Design requirements for earthing and protection in island mode are discussed in Section 9.4.

Section 6 – Other EESS components

6.7   Monitoring and metering

Most EESS will feature monitoring and metering. These may be stand-alone units or devices that are accessible over the web or that post data to a web portal.

Parameters that may be monitored by an EESS monitoring system include:

(a) battery state of charge;
(b) battery voltage;
(c) battery charge/discharge current/power;
(d) battery temperatures;
(e) output power, voltage and frequency;
(f) ambient and EESS equipment temperatures;
(g) grid voltage and frequency;
(h) error logs; and
(i) calls for maintenance actions.

Most EESS will feature a user interface to provide monitoring and metering. This may be one or more stand-alone units, integrated graphic display screens, or devices that are accessible on a mobile app via the internet or that post data to a web portal.

6.8   Grid synchronization

Synchronization is the process of minimizing the voltage, frequency and phase differences between a generator and a powered system (such as the grid) to which it will connect.

To connect to the grid, a self-commutated inverter operating in island mode adjusts its voltage and frequency using its automatic voltage regulation and automatic frequency governor controls, to bring it within an acceptable range before the final contactor or circuit-breaker is closed to make connection with the grid.

 Section 7 N

EESS safety and planning considerations

7.1 Planning

Whether planning permission is required for the installation and operation of an EESS will depend on:

(a) the nature, scale and location of the EESS and its associated housing and transmission infrastructure; and

(b) the relationship between the EESS and the existing use of the land and buildings in which it is located.

In addition, the Building Regulations of the relevant part of the UK will need to be considered.

The installer shall make clear who is responsible for obtaining planning permission where it is necessary. This is usually the householder, for domestic customers.

If the EESS is housed in an existing building, is associated with the existing land use and is not of a scale likely to give rise to a material change of use or breach of existing planning condition, planning permission is unlikely to be required. It would be likely regarded as the installation and operation of plant and machinery requiring only internal, non-structural alterations and a use related to the main use of the land.

Larger scale EESS requiring the erection of external housing (whether free-standing or as an extension to an existing building) are likely to constitute development, requiring planning permission, although an application for express planning permission may not be required in every case.

7.2 Safety

This section provides a suggested list of safety legislation, standards and considerations that designers, installers and maintainers of EESS will encounter. Necessarily, this section cannot be exhaustive, and it is the responsibility of the individual designer, installer and maintainer to determine the applicable legislation for a particular installation or operation, and adopt suitable standards where appropriate.

7.2.1 Relevant legislation

Table 7.1 provides a commentary on some key legislation applicable to the provision of most EESS. Table 7.2 provides a commentary on further legislation that may apply in places of work.

These lists are not exhaustive and further legislation, or later amendments, may apply. The reader should ensure they comply with all relevant legislation.

Section 7 – EESS safety and planning considerations

Table 7.1 Legislation relevant to the provision of EESS

Legislation	Notes
Electrical Equipment (Safety) Regulations 2016	Requires any supplied electrical equipment to be safe.
	This is relevant to the procurement of electrical components and the supply of those, particularly (but not limited to) domestic customers.
	Implements the Low Voltage Directive in the UK.
Electricity Safety, Quality and Continuity Regulations 2002	Core legislation relating to public supplies. Contains requirements for public suppliers and also certain requirements for consumer's installations.
	There is a particular requirement in 'PART VI' to agree specific requirements or notify the distributor operating the network to which the EESS may be connected.
Electromagnetic Compatibility Regulations 2016	Requirements for the electromagnetic compatibility of equipment and installations.
	Relevant to the procurement, installation design, erection and maintenance of EESS.
Waste Batteries and Accumulators Regulations 2009	Provisions for safe and environmentally friendly disposal of batteries and accumulators. Principal duties are for manufacturers and importers.
	This is relevant to the procurement, supply and disposal of storage batteries.
Electricity at Work Regulations 1989	Covers the requirements for working safely with electrical installations and equipment and also the basic requirements for the safety of electrical equipment and installations in places of work.
	This is relevant to the design of the EESS for safety in installation, commissioning, maintenance and decommissioning, and for the safety of EESS used in places of work.
	The Electricity at Work Regulations apply to electricians working on installations and equipment in dwellings.
Health and Safety (Signs and Signals) Regulations 1996	Provisions for safety signs and safety signals. This is particularly relevant for safety signs selected for use in, or with, the EESS.
The Construction (Design and Management) Regulations 2015	Known as the 'CDM Regulations', they provide the main regulations for managing the health, safety and welfare of construction projects.
	Construction projects above a certain size (measured in time or staffing levels) must be notified in advance.
	There are specific duties for clients, designers, and contractors in the design and execution of the project. Some of these duties apply regardless of whether the project is a notifiable project.
	Commercial and business clients should appoint in writing both a Great Britain-based principal designer and a principal contractor, otherwise the client themselves is deemed to take on those roles.
	In private domestic work, the principal contractor assumes some duties otherwise assigned to the client.
Responsibility for building regulations throughout the UK is a devolved matter: England & Wales – Building Regulations 2010, as amended (note that some amendments in Wales are different to those in England); Scotland – Building (Scotland) Regulations 2004, as amended.	The requirements of the various Building Regulations cover safety and environmental provisions amongst other things. They also provide a system whereby applicable work is notified to the relevant local building control body. Further guidance is available in the relevant Approved Documents (England & Wales) and Technical Standards (Scotland).
	The provision of a new EESS will often involve the provision of new circuits or the changing of protective devices, and in these cases the work is generally notifiable. See the relevant Approved Documents (England & Wales) and Technical Standards (Scotland) for further details.

Section 7 – EESS safety and planning considerations

Table 7.2   Additional legislation relevant to the provision of EESS in places of work

Legislation	Notes
Management of Health and Safety at Work Regulations 1999	Reinforcing the Health and Safety at Work etc. Act 1974, these regulations outline how health and safety at work is to be managed.
The Dangerous Substances and Explosive Atmospheres Regulations 2002	Requirements and responsibilities for identifying and managing the risks from fire, explosion and corrosion of metal.
Regulatory Reform (Fire Safety) Order 2005	Outlines the responsibilities for organizations to manage the fire safety of premises, identifying and managing fire safety risks.
Control of Electromagnetic Fields at Work Regulations 2016	Requirements for identifying risks associated with electromagnetic fields in the workplace and managing them. Action levels and exposure limits are regulated.
Confined Spaces Regulations 2009	Legislation identifying responsibility to assess and manage risks associated with confined spaces. A confined space is a place that is substantially (though not always entirely) enclosed and where serious Injury can occur from hazardous substances or conditions within the space or nearby (e.g. lack of oxygen).
The Dangerous Substances (Notification and Marking of Sites) Regulations 1990	This legislation requires those in control of a site or premises where more than 25 tonnes of 'Dangerous Goods', as classified by the Control of Dangerous Goods Regulations, are used or stored, to give written notice to the Fire Service and the Health & Safety Executive. This includes lithium-ion batteries (UN 3480), and lithium-ion batteries in or with equipment (UN 3481), and would therefore impact the planning for large EESS using this type of storage.

7.2.2   Relevant standards

See Appendix D.

7.2.3   Risk assessment

Hazards should be considered for all stages of the lifecycle (design and planning, transportation, installation, commissioning, operation, maintenance, repair and end-of-service life).

All system components should be designed, manufactured and tested in accordance with relevant safety standards.

The integration of EESS components should be considered within the risk assessment. In particular, incompatibilities between components should be considered.

The assessment should consider the total energy stored in the EESS and population that might be affected by particular hazards.

7.2.4   Key considerations – hazard inventory

The inventory presented in Table 7.3 is intended to outline general considerations for designers, installers, operators and maintainers of EESS. It does not represent all hazards that may be present in a particular installation or for a particular application or operation. Further, detailed, information on designer's considerations for risk assessments are presented in the Construction Industry Research and Information Association (CIRIA) guidance (C755 and C756).

Section 7 – EESS safety and planning considerations

Table 7.3   Hazard inventory

Hazard/issue	Notes
Safe isolation	In order that an electrical system can be safely maintained, and to ensure that energy can be disconnected in a fault or for emergency purposes, it is important to consider isolation as part of the electrical design of an EESS. EESS have more than one source of electrical energy that can potentially deliver hazardous voltages and currents. Schematics and isolation instructions must clearly identify appropriate isolation procedures to ensure safety and adequate warning signs and notices must be provided. The electrical design of an EESS is discussed further in Section 9.
Protection against electric shock	Protection against electric shock for all modes of operation must be in accordance with BS 7671. In particular, the need for earthing and additional protection is important when operating in island mode. In this mode, the supplier's earth (if it was available on-grid) may not be available, and the prospective fault currents from an inverter are much less than those available from a grid-supply. The electrical design of an EESS is discussed further in Section 9.
Fault current protection	As with protection against electric shock, adequate fault current protection must be in place for island mode operation due to the lower prospective fault currents from inverters. The electrical design of an EESS is discussed further in Section 9.
Weight of components	Batteries are typically heavy devices. They should be located at a suitable height for lifting in and out of their housings for maintenance replacement. Surfaces (walls and floors) must be capable of supporting the weight of any components mounted on them. Installation considerations are discussed in Section 11.
Batteries	Batteries pose electrical hazards because of their high prospective short circuit currents. Adequate fault current protection must be in place. Additional hazards associated with a battery may include: • chemical hazards (e.g. caustic, corrosive or toxic chemicals); • charging hazards (thermal considerations, evolution of gases, cell reversal, etc.); and • electrical hazards (batteries are a continual source of power that cannot be fully isolated, and safe systems of work are therefore often required for fault-finding, maintenance, replacement, commissioning, decommissioning, and preparation for transportation of batteries). The battery manufacturer should always be consulted in relation to hazards and requirements for safety in transportation, storage and use. Such information, along with assessments of DC arc flash risk where appropriate (see Appendix E), must be made available to users and maintainers. Batteries are discussed further in Section 5.
Environmental conditions (temperature and ventilation), including fire and explosion risks due to electrical hazards, or explosive gases	Certain battery types are susceptible to damage, or even thermal runaway, if the temperature is allowed to escalate during charging. Many lithium-ion cells designed for use in EESS contain in-built safety monitoring to prevent thermal runaway. Certain battery technologies, such as lead-acid and valve regulated lead acid (VRLA) types, require fresh air ventilation to prevent accumulation of hydrogen in explosive concentrations. Hydrogen may be produced during faults on the battery or charger. The air refresh rate should be calculated in line with the guidance of BS EN IEC 62485 and any further information provided by manufacturers. Installation considerations are discussed in Section 11.

Table 7.3   *Cont.*

Hazard/issue	Notes
Fire safety, firefighting and fire suppression	The fire risk assessment for the premises should be consulted and, where necessary, revisited, when designing and installing an EESS. Fire suppression technologies in use in the location, or recommended by the fire risk assessment, must be compatible with all of the EESS components. The selection of inappropriate firefighting equipment or fire suppression systems for a particular battery type may have devastating results. Installation considerations are discussed in Section 11.
Flooding or water pipe burst	A site Flood Risk Assessment may need to be carried out or updated. Consideration may be required of hazards relating to the storage becoming wet during flooding, pipe burst, etc.

Section 8 N

Specification of an EESS

8.1 System specification

The selection and sizing of an EESS depends on the application. While determining the power and energy requirements of the EESS is obviously necessary, understanding the intended operating mode (or combination of intended modes) is equally important. For example, a system sized to provide a solar time shifting function (increased self-consumption) may be significantly smaller than the same system that also needs to provide energy during power cuts. Similarly, a system intended to provide frequency response services (FFR) may require less storage capacity than one intended to provide both FFR and peak avoidance.

Considerations for selecting and sizing an EESS for some of the more common operating modes are presented in this section.

Where an EESS is installed into an electrical system featuring embedded generation (such as an existing PV system), compatibility between the EESS and on-site generation needs to be verified for all operational modes.

8.1.1 Modelling

Some degree of system modelling will be required for all applications. The extent and complexity of the modelling will vary depending on the scale of the system and the application. For large-scale projects providing grid services, detailed modelling of power flows and battery state-of-charge over short data intervals throughout the day and year can be expected.

Factors that may be incorporated in modelling are shown in Table 8.1.

Table 8.1   Factors that may be incorporated in modelling

Factor	Relevance
• Renewable system size (e.g. PV array kWp). • Renewable system daily and monthly generation profile. • Building load profile.	Particularly relevant for systems providing time shifting function (increased self-consumption).
• Site daily energy use. • Site peak power requirement. • Load characteristics (e.g. identification of any loads with high start-up requirements).	Particularly relevant for systems providing a back-up/island mode/grid-independent power supply.
• Site load profile (e.g. from half hourly meter data). • Site power factor data. • On-site generation patterns (renewables, CHPs etc.). • Grid voltage & frequency data. • Energy bill data (time of use pricing etc.). • Site import/export constraints.	Particularly relevant for systems providing grid services, peak avoidance, energy arbitrage etc. Load profile data at a higher resolution of 5-minute data may be required to fully evaluate the benefits of storage in dwellings, due to the volatility of their energy demand profiles.

A variety of computer tools, including those from manufacturers, are available to help model system performance and enable the selection of appropriate EESS characteristics. Some may be EESS specific tools while others may be a function of other software (for example, proprietary PV software is commonly used to model EESS that is designed to add a time shifting function to a solar PV installation).

Section 8 – Specification of an EESS

8.1.2 Load profiles

Most applications will utilize some form of load profile in the modelling process.

Load profiles vary seasonally, and often in a manner that does not favour self-consumption of the installation.

Generic load profiles for common sites are available for use. These vary in sophistication but, for example, a typical modelling software package may allow the user to select between different types of domestic property (for example, two-person household working during weekdays, two-person household retired etc.). When using a standard profile, the system designer needs to ensure the relevance of the profile selected – for example, ensuring that it is a profile for a UK home, whether any degree of electric heating is assumed etc.

However, while useful, generic load profiles can vary considerably from the load use on the particular site in question. Most modelling software packages allow a site-specific load profile to be uploaded. Such site-specific data can be obtained from various sources, for example, from:

(a) half-hourly electricity bill records;
(b) existing building monitoring systems; and
(c) metering installed specifically to capture data for the purposes of designing an EESS.

Where site data is utilized or where metering is specifically installed to capture data to facilitate the design of an EESS, the designer needs to ensure that the data collection period is long enough and covers a sufficiently representative period of the site's operation over the year.

The data set also needs to suit the intended application of the EESS. For example, where an EESS will be providing electricity during power cuts, the data recording needs to capture the instantaneous peaks – data from 15 minute averages, for example, may not give an accurate representation of the peak loads that the system will need to supply.

For some sites, high resolution data is needed to predict system behaviour. For example, domestic customers would typically require 5-minute or higher resolution data to accurately predict self-consumption rates.

8.2 Sizing an EESS for renewable time shifting (maximizing self-use)

There is a common perception when selecting an EESS that bigger is always better. However, for an EESS designed to maximize self-use of a grid-connected renewable energy system this is not necessarily the case. Sizing an EESS for a renewable time shifting application depends on the nature of the site and the intended use.

Section 8 – Specification of an EESS

8.2.1 Capacity (kWh/MWh)

Selecting the appropriate capacity is often an iterative process achieved by trying different options in the modelling software. The optimum EESS capacity (kWh/MWh) lies somewhere along a continuum between two extremes, as illustrated in Figure 8.1.

Figure 8.1 Effect of extremes of EESS capacity

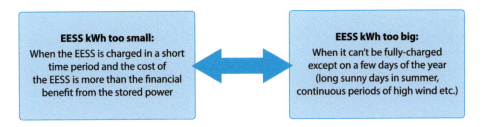

8.2.2 Charge/discharge capability (kW)

EESS generally have limits on the amount of power they can handle (charge/discharge) at any one time. This may be due to an inherent limit in the EESS technology, battery size and technology, or in other components of the system (for example, an inverter).

Again, bigger isn't necessarily better for all characteristics. For example, due to the pattern of load use and/or renewable generation, many systems may operate for a significant period of time at low charge-discharge rates. Selecting an EESS or inverter optimized for the majority of the intended operation (rather than the peak) can have both cost and system efficiency benefits.

Figure 8.2 Example chart showing proportion of total hours spent at a discharge power throughout a year

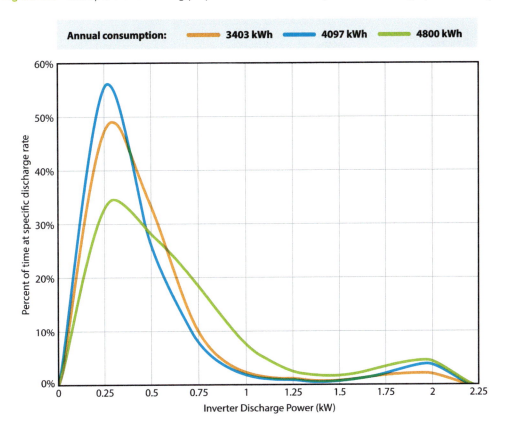

Section 8 – Specification of an EESS

8.2.3 Site and system factors

Even when using modelling software, system selection is often an iterative process with the designer seeing how well different options perform in the model. This process can be made quicker if some options can be excluded (for example, by determining the maximum useful kWh).

Factors to consider when performing the initial scoping of an EESS for use in a renewable time shifting (maximizing self-use) application include the following.

(a) Renewable system output
As the EESS capacity increases, there comes a point where the daily renewable input becomes insufficient to fully recharge the battery. Although renewable output varies considerably during the year, the maximum daily output is relatively easy to determine and can be used to help fix an upper battery threshold.

For systems that don't provide any island mode, DC battery backup or UPS/CPS functionality, the upper storage capacity to consider during calculations is likely to be around 1-2 days of storage. For systems providing island mode, DC battery backup or UPS/CPS functionality, it is likely to be more, for example, 5-7 days.

Figure 8.3 illustrates the effect of capacity where an EESS is coupled to a solar PV system. Where the EESS capacity (kWh) is sufficiently large enough, all of the daily solar excess can be stored. Systems with a relatively smaller battery will not be able to store all the excess electricity.

Figure 8.3 Effect of capacity where EESS is coupled to a solar PV system © IET 2017 (Reproduced as adapted from *Battery energy storage systems with Grid-connected solar photovoltaics – A Technical Guide (BR 514);* © IHS Markit 2017)

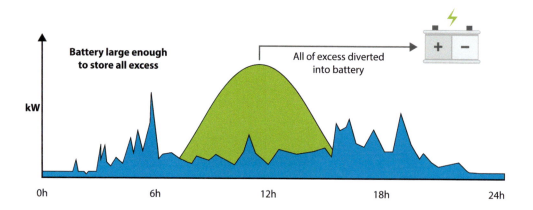

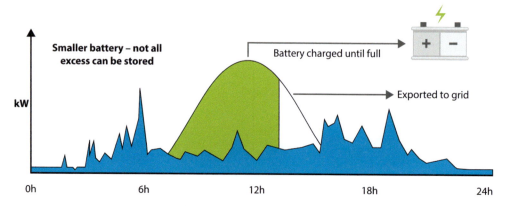

Note: While the above diagrams are a useful way of illustrating the impact of battery size, real world considerations are more complex as the solar excess varies very significantly during the day (as solar and load conditions change), across the year and from site to site.

(b) Load use

Load use and character varies considerably from site to site. Sites with loads that run throughout the day will reduce the available renewable power for battery recharging and mean that a smaller battery is generally required.

(c) Winter mode

Where systems feature a winter sleep mode (particularly relevant for a solar EESS), system operation for the period of the year when the system will be put into winter sleep mode can typically be ignored during the modelling process.

Note: Significant amounts of time without any charge/discharge cycle can reduce overall life and capacity of a battery system.

(d) Back-up capacity

Where power is required during a power cut, some or all of the battery capacity can be set aside for these occasions. As noted above, this will generally mean that a significantly larger battery is required than one specified solely for daily time-shift mode operation. More information for sizing systems to provide island mode or back-up power is presented in Section 8.3 below.

(e) Maximum charge/discharge rate

For some systems, the maximum charge/discharge rate that the EESS is able to handle will be relevant. For example, an EESS selected for island mode functionality needs to be able to deliver the worst case (surge) currents that the inverter will require during island mode operation.

8.3 Sizing an EESS for back-up and island mode systems

Where the back-up capability of the EESS is a key part of the design, a load evaluation needs to be performed. This needs to include an examination of the power rating of the load(s) to be run (watts) and how long they will need to run for (hours). These can then be multiplied together to provide the energy required (watt-hours).

For many sites it will not be appropriate to try to run all the loads from the EESS. In these circumstances a selection of loads ('critical loads') are electrically separated from the rest of the loads, enabling just these loads to be run from the EESS during a power outage. This approach allows a smaller EESS to be specified.

Note: Name-plate ratings may not provide accurate information on all loads. For some loads the declared figure may be the peak or surge rating, hence actual continuous power use may be less. For more accurate information, load power use can be measured using an appropriate meter.

An example of how to evaluate the load requirements is illustrated in Table 8.2.

Table 8.2 Example of load evaluation

Item	Run time (hours)	Rating (W)	Energy (Wh)
Laptop	2	50	100
Light	2	25	50
	Subtotal	75	150
	Multiplier to allow for system losses	1.1	
	Number of days' storage required	5	
	Energy total (Wh)	825	

Note: A multiplier is used here to allow for system losses. The system designer needs to select an appropriate value for the system, but it is generally recommended that a figure of at least 1.1 is used as the multiplier.

Energy total (Wh): this figure is used to help determine the energy storage capacity of the EESS required for island mode operation.

Power total (W): this figure is used to help determine the size of inverter needed in the EESS. However, in practice, it may not be necessary to size the inverter to the maximum figure calculated here as a degree of diversity can be assumed (not all loads on at the same time).

In addition to the inverter power calculated here, the start-up power requirement of the loads needs to be considered and compared to the capability of the inverter. For sites with loads that have significant start-up requirements (for example, motors) an increased inverter capacity may be required.

Loads such as motors (including those in fridges and freezers) and electronic power supplies (including lighting and information and communication technology equipment) will need particular investigation. It should also be considered that, following switchover to EESS operation, it is likely that all loads will re-start simultaneously.

Techniques to help reduce the impact of switch-on or switch-over inrush currents include:

(a) zero-voltage or zero-current switching, where loads are re-energized as the voltage, or current, respectively, approaches zero;
(b) direct on line starting for motor loads; and
(c) delayed, or staggered, start, where loads are brought on in stages to alleviate the rapid differential in demand experienced by the inverter.

> **Preventing inverter overload/premature battery drain**
>
> To prevent users from inadvertently overloading an EESS during island-mode operation, a label showing the maximum capacity of the system may be appropriate. This is of particular relevance where the circuits provided during a power-cut include sockets accessible to the user.
>
> Careful selection of circuit protective devices (such as MCCBs) can also help to ensure that the inverter is not overloaded during island-mode operation. Where appropriate, this may include – for the circuits provided during a power cut – the use of circuit protective devices with a lower rating than would otherwise normally be specified for that circuit.
>
> Ensure that user awareness of the limitations of island-mode operations is adequately covered in the operation and maintenance (O&M) manuals and handover process (see Section 13).
>
> A system to monitor the load on the inverter and give a visual and/or audible warnings when approaching or exceeding the inverter's capacity might also be useful in some circumstances.

8.4 AC coupled system design – component compatibility check

With AC coupled systems, the compatibility of all parts of the system needs to be verified. In particular, a check needs to be made to ensure that components of the system are not at risk of damage during island mode operation.

Section 8 – Specification of an EESS

AC coupled systems are commonly installed as a retrofit to an existing renewable system. In such cases it is important that the installer ensures that the existing renewable system inverter(s) is compatible with the new equipment being installed. Similarly, the new EESS needs to be checked to ensure compatibility with the existing system components.

Where an AC coupled design features an isolation relay on the grid supply (to enable island mode operation), the existing renewable system will still 'see' an electrical supply during a power cut (as provided by the EESS). Hence, the renewable system will attempt to resume generation and feed electricity into the 'islanded' system. All parts of the system need to be checked to ensure that they are compatible with this island mode operation; or isolated during island mode. Where the renewable system remains operating, a means to curtail output when the battery is fully charged needs to be implemented (via direct communication between the EESS and the generator or via other means such as frequency droop control).

Figure 8.4 AC coupled EESS with isolator on existing generation

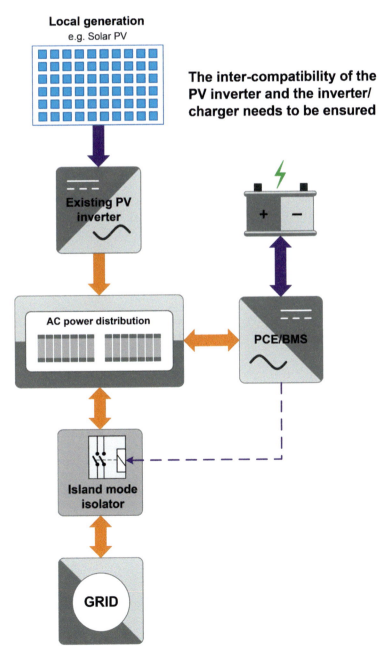

Section 8 – Specification of an EESS

For some systems designed to work in island mode and configured as shown here, the EESS may need to have a power capacity that is greater than that of the renewable system.

In an island mode scenario there is the possibility that generation could exceed load demand. Where this excess occurs suddenly (e.g. where a load is switched off), there will be a period where there is excess generation with nowhere for the energy to go, resulting in a short-term overvoltage. If the system cannot accommodate this excess generation, then there is the possibility of a component failure due to overvoltage.

This scenario should be checked with the manufacturers of all the equipment involved.

In some cases, products may be able to handle this scenario. In other cases, the issue may be mitigated by arranging for the EESS to absorb the excess generation into its battery; this may require the EESS to have a charge rate similar to or exceeding that of the on-site renewable generator and have some spare battery capacity.

Note: While increasing EESS power rating may solve the issue, similar problems can occur once the EESS is fully charged (the EESS may not have the capacity to store the excess generation) and this scenario also needs checking with the manufacturers.

Where an EESS is installed into an electrical system featuring embedded generation, the inter-compatibility of the EESS and on-site generation needs to be verified. In particular, a check needs to be made to ensure that no problems can occur when an EESS and on-site generator are both feeding the same islanded electrical sub-system.

Where compatibility cannot be verified, the local generator can be connected to the grid via an isolation relay – linked to the main change-over relay. This will disconnect the generator from the rest of the system during a power cut.

Figure 8.5 AC coupled EESS with isolator on existing generation

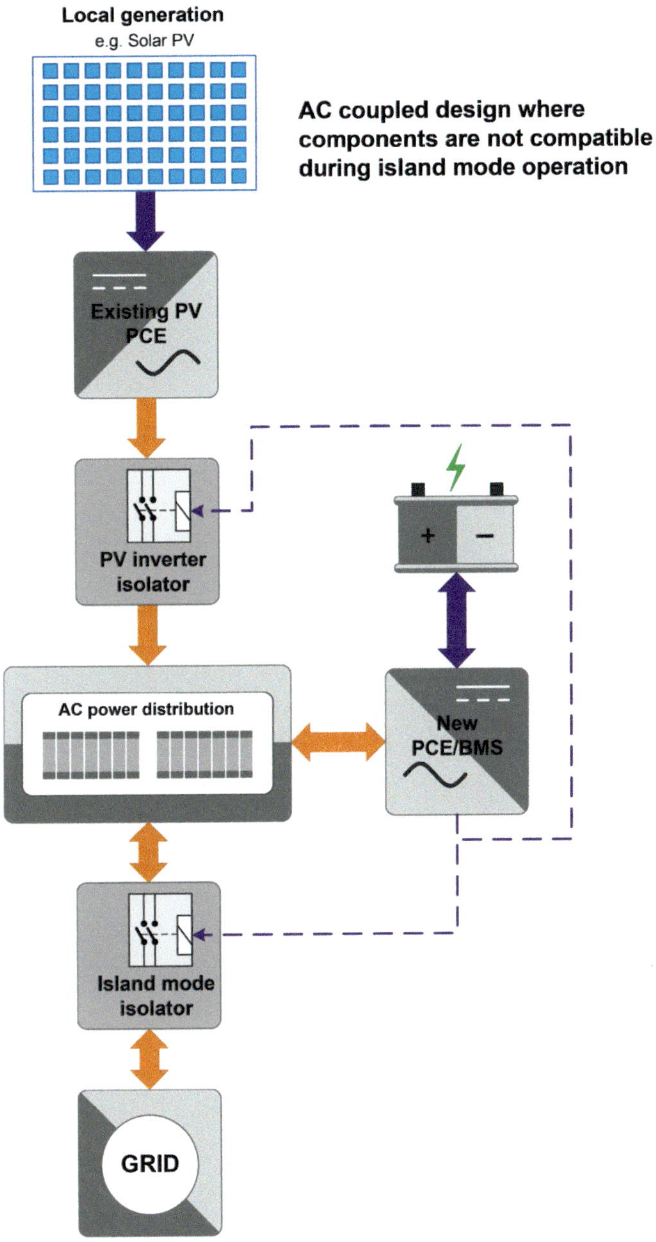

8.5 Sizing an EESS for grid services

Sizing an EESS to provide grid services generally requires detailed consideration of the requirements of the network operator relating to the intended service or combination of intended services.

For example, when sizing a system intended to provide frequency response services, grid frequency records can be used to determine how often, how long and at what power levels the EESS may be expected to operate on a day-by-day and month-by-month basis. This operational data can then be used to analyse the storage capacity required by the proposed EESS. It can also be used to assess the daily cycling and DOD that will be required from the EESS – which can then be used to select an EESS with a suitable capacity and cycle warranty.

▬ Section 9   N

Design of an EESS

This section outlines the key electrical installation design considerations for an EESS for LV installations.

9.1   General consideration of electrical installation standards

9.1.1   BS 7671 (IET Wiring Regulations) and BS HD 60364-8-2

BS 7671 is the principal electrical installation standard and applies to LV and ELV installations, and the selection and erection of equipment. Certain key parts of the standard require careful consideration for EESS installations, and these are highlighted throughout Section 9 of this Code of Practice.

BS HD 60364-8-2 covers functional requirements for prosumers' electrical installations, and is therefore directly relevant to the functional design of an EESS. This standard may be considered for future inclusion into BS 7671 as Chapter 82.

9.1.2   ENA Engineering Recommendations G98, G99 and G100

The requirements of these Engineering Recommendations are discussed in detail in Section 10. They relate not only to the process to be adopted for connection but also to the technical requirements (such as the requirement for protection in accordance with G99 for larger systems). ENA Engineering Recommendation G100 addresses export-limiting systems.

Note: G99 replaced G59 and G98 replaced G83. G98 and G99 were applicable from 27 April 2019.

Table 9.1   Summary of the requirements of the ESQCR and ENA Engineering Recommendations G98, G99 and G100 for various arrangements of EESS

Arrangement of EESS	Relevant ESQCR Regulation for EESS	Relevant ESQCR Regulation for EESS Applicability of ENA Engineering Recommendations G98, G99 and G100
An EESS that can charge from the grid, but has inverter outputs that are never connected in parallel to the grid. Examples include: (a) UPS, CPS, or other generators that only operate as switched alternatives to the mains supply; (b) DC battery backup type systems as illustrated in Figure 3.3; and (c) other types of EESS installed in this particular way that never connect parallel with the grid.	21	ESQCR Regulation 21 requires the installation to comply with the requirements of BS 7671, and provides that the equipment or installation must be arranged so the generation sources are never connected in parallel with the grid.
EESS capable of parallel operation (both EESS and grid feed loads in parallel).	22	ENA Engineering Recommendations G98 or G99 apply.
Export-capable EESS.	22	ENA Engineering Recommendations G98 or G99 apply. ENA Engineering Recommendation G100 may apply where export limiting is required. Must be installed so it can't charge from grid, then claim export tariff for this charge later.

Table 9.1 *Cont.*

Arrangement of EESS	Relevant ESQCR Regulation for EESS	Relevant ESQCR Regulation for EESS Applicability of ENA Engineering Recommendations G98, G99 and G100
Storage is DC coupled to an existing grid-forming inverter (e.g. for solar PV).	n/a	Must be installed so it can't charge from grid, then claim export tariff for this charge later. **Note:** The existing inverter itself is subject to the requirements of Regulation 22 of the ESQCR, and ENA Engineering Recommendations G98, G99 and G100 as appropriate.
EESS is used in a wholly grid-independent system such as those illustrated in Figure 3.10 and Figure 3.11.	n/a	n/a

9.1.3   BS EN 50171

BS EN 50171 applies to the design and installation of CPS for emergency lighting and similar applications, and therefore is typically not relevant to the design of domestic EESS. Should a commercial or industrial installation require the EESS to perform functions of, or integrate with, a CPS, BS EN 50171 will be relevant.

9.1.4   BS EN IEC 62040

The BS EN IEC 62040 series of standards applies to the design of UPS. If the EESS is required to provide UPS functionality, relevant parts of BS EN IEC 62040 will apply.

9.1.5   BS EN IEC 62485

BS EN IEC 62485-1 provides general safety requirements for LV secondary (rechargeable) battery installations. BS EN IEC 62485-2 provides additional requirements for installations of stationary batteries for lead-acid and Ni-Cd battery types.

This would be used by the installation designer to consider safe location and ventilation of pre-manufactured battery enclosures, or to design suitable battery enclosures or rooms for a discrete component or large commercial or industrial system.

9.2   The EESS battery and PCE/BMS DC arrangements

9.2.1   General

The principal risks associated with the electrical installation for battery, PCE and BMS are:

(a) many batteries have a low internal resistance and, consequently, the prospective short circuit currents can be very high.

Especially with higher amounts of stored energy, arc flash can be a distinct issue resulting from available short circuit current. Arc flash hazard studies should be performed in accordance with a reputable methodology for calculating arc flash. See Appendix E.

(b) unprotected short circuits may lead to the battery overheating and, if unprotected, this overheating can damage the battery. This can happen with any type of large battery. Some batteries require fast disconnection times to prevent thermal runaway and damage, and therefore an appropriate overcurrent protective device should be selected based on the battery chemistry as well as achieving BS 7671 requirements for fault current and overcurrent protection.

Section 9 – Design of an EESS

(c) DC circuits cannot be protected by commonly available RCDs. Other means of additional protection should therefore be considered where relevant. Where batteries are located in a separate enclosure to the charger and/or inverter, the use of protective (metallic, earthed) containment, armoured or mineral insulated cables should be considered. An alternative may be double or reinforced insulation (for example, sheathed cables) in a wiring support system offering equivalent mechanical protection.

(d) means of isolation needs to be provided for battery replacement/maintenance.

(e) battery packs or cells cannot (in general) be individually made dead, and a safe system of work is required to minimize risks of injury from accidental contact or short circuit.

Figure 9.1 shows examples of the battery electrical connection arrangements (centre-tapped earth, positive earth and negative earth) with battery and charger/inverter located in separate enclosures. Figure 9.2 illustrates a possible arrangement where the battery and charger/inverter are in the same enclosure.

Note: The arrangement of the battery and charging equipment must take into account clearances required for DC arc flash (see Appendix E), in addition to considering any separation requirements for chemical or gas explosion risks.

In larger systems, the batteries may be arranged in battery rooms or large outdoor enclosures, and presented as open-frame arrangements when the rooms or enclosures are opened. Access to such locations should be for authorized and suitably trained/competent personnel only, and secured by lock and key (or tool access for smaller enclosures). Where the EESS is to be installed in a dwelling or smaller location, access to individual cells or cell packs should be only via tamper-resistant fastenings operated by a suitable tool.

In general, the following are recommended:

(a) protection against electric shock for the battery should be by one or more of the following methods:
  (i) where the battery and charger/inverter are in separate enclosures (see Figure 9.1), use one or more of:
    • SELV;
    • PELV;
    • automatic disconnection of supply (provided by earthing and appropriate fuse or other suitably rated overcurrent protective device); and
    • double insulation or reinforced insulation, provided mechanical protection is suitable.

Note 1: Where the battery and charger/inverter are in separate enclosures, in order to protect against thermal effects and arcing due to undetected faults, functional and/or protective earthing is strongly recommended, regardless of voltage. Mechanical protection should be provided for interconnecting cables.

Note 2: The temperature of batteries should always be monitored in the battery enclosure. In addition to power cables, control and monitoring cables, such as battery temperature monitoring cables, that are connected between charger/inverter enclosures and battery enclosures, are key to battery safety, and shall also have suitable mechanical protection. In some systems, it may be necessary to treat control and monitoring circuits as circuits of safety services.

  (ii) where the battery and charger/inverter are in the same enclosure (see Figure 9.2), use one or more of:
    • SELV;
    • PELV;
    • electrical separation; and
    • automatic disconnection of supply.

Note: Commonly available RCDs cannot be used with DC supplies, and as a result, the use of a TT earthing arrangement for battery circuits for smaller installations is not usually practicable.

(b) where not achieved by other means (for example, current limiting by design or internal protection), the battery output should be protected by an overcurrent protective device, located as close as practicable to the battery.

(c) uninsulated conductors of battery circuits at any voltage, other than protective conductors, should only be accessible using a key or tool, since even where the shock risk is minimal with low values of ELV, thermal and arcing hazards remain.

Section 9 – Design of an EESS

(d) overload and short-circuit protection is essential for all battery voltages. Circuits in EESS typically experience bi-directional current flow. Careful selection of overcurrent protective devices is required, as not all of them are rated for bi-directional current flow.

(e) persons should be warned of the hazard of battery voltages in excess of 60 V DC by warning signs (see Clause 11.1 of BS EN IEC 62485-2).

(f) battery cells or cell packs should be installed so that they are inaccessible to ordinary persons, i.e. in a secure location or secure enclosure for smaller batteries.

(g) battery terminals should be arranged so that two bare conductive parts, having between them a voltage exceeding 120 V, cannot be inadvertently touched simultaneously. Methods of achieving this in accordance with BS 7671 include fitting battery isolation covers, or the use of appropriately arranged barriers.

(h) safe systems of work should be in place for fault finding, maintenance, commissioning and decommissioning activities associated with battery packs/cells.

(i) a battery isolator is provided (strongly recommended). The isolator should be located physically and electrically as close as possible to the battery, after or forming part of the battery fault current protection. In order to help prevent accidental short circuit and electric shock during maintenance, double-pole isolation is recommended for battery voltages above 48 V DC and is essential for LV batteries (above 120 V DC). The means of isolation should be selected in accordance with BS 7671. Where the battery and PCE are located in the same sealed enclosure, and maintenance of the storage battery on site is not envisaged, an isolator for the storage battery is not always necessary.

An on-load isolator should be specified where the battery may be replaced/maintained with the EESS controller still operational. Certain plug and socket-outlet combinations are not suitable for on-load isolation. Where the EESS controller is to be isolated before battery maintenance/replacement, a plug and socket-outlet arrangement may be suitable.

Section 9 – Design of an EESS

Figure 9.1 Example of EESS battery and PCE in separate enclosures

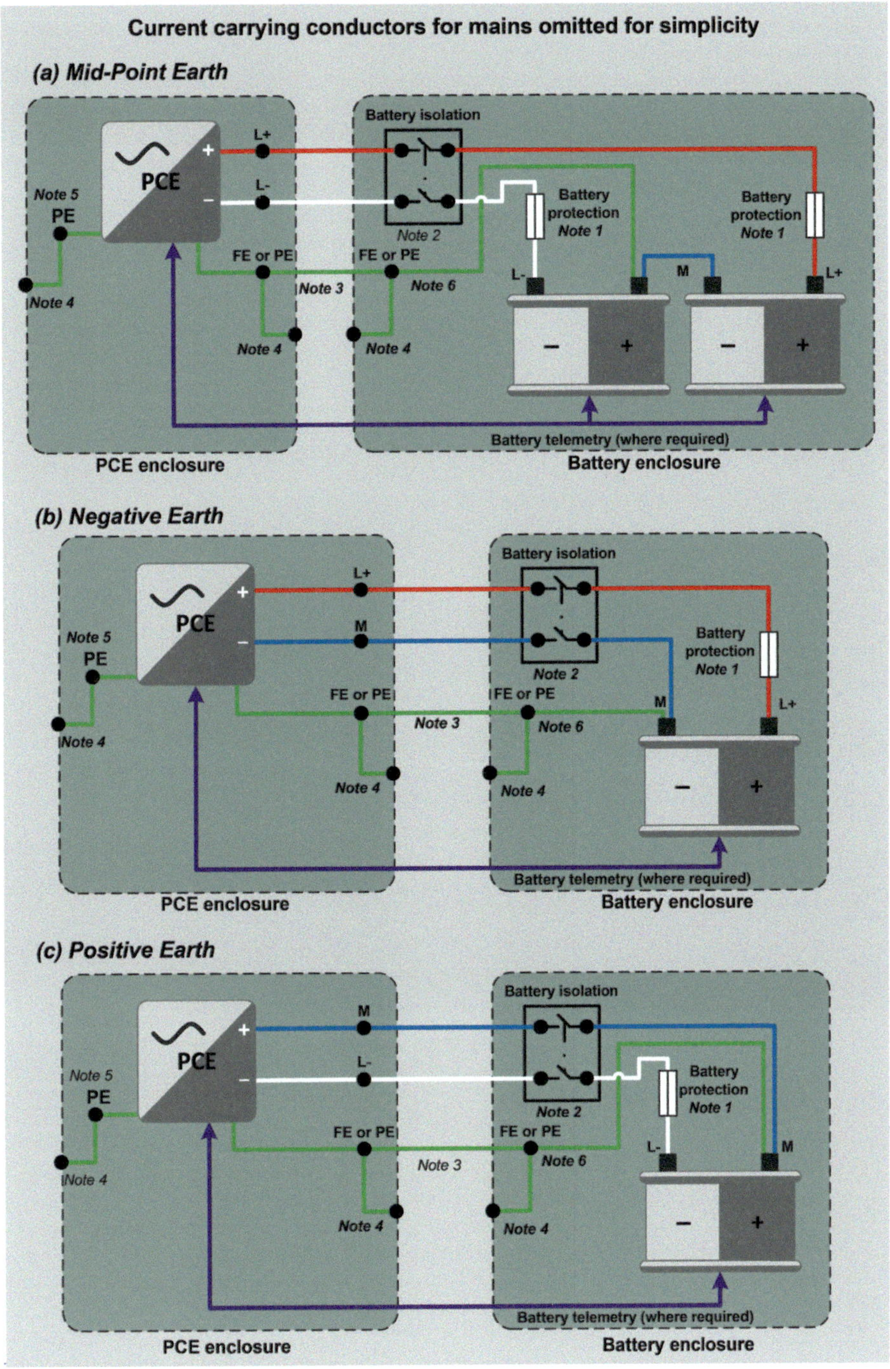

Notes to Figure 9.1

Note 1: Where required, see Section 9.2.1(a). To be located physically and electrically as close as possible to the battery. Must be rated for DC protection, and be capable of breaking at least the maximum short-circuit current of the battery. Whilst a fuse is shown, an MCB or MCCB, rated for DC protection, is an equally acceptable alternative.

Note 2: Should be rated as an on-load isolator. To be located physically and electrically as close as possible to the battery, after the battery protection fuse (see Note 1). Double-pole isolation is recommended for battery voltages above 48 V DC and is essential for LV batteries (above 120 V DC). The isolator may not necessarily be in the battery enclosure. Where the battery and PCE are located in the same pre-manufactured enclosure accessible only by the use of a tool, and maintenance of the storage battery on site is not envisaged, an isolator for the storage battery is not always necessary.

Section 9 – Design of an EESS

Note 3: Earthing for the DC battery circuitry is used to operate protection in the event of a fault, i.e. for functional purposes. It may not be required for protection against electric shock where ELV is used and relevant conditions of BS 7671 for FELV, PELV or SELV are maintained in the DC circuits. It is preferred that the cable between the battery enclosure and EESS enclosure is suitably protected against mechanical damage – metallic conduit, trunking, metallic armoured cable, or MICC may be suitable for this purpose.

Note 4: Connections of earthing to conductive enclosures or conductive parts of battery frames, racks or stands.

Note 5: LV protective conductor. Ideally, this would be solidly connected to the DC functional earth (see Note 3); however, this may depend on the electronic design of the PCE, and the PCE manufacturer's instructions should be consulted.

Note 6: It is preferred that the battery is earthed, even if operating at ELV, so that a fault in interconnecting cable connecting the battery can be detected. For LV battery arrangements, earthing is required. Battery earthing (and wiring colours) are illustrated in a negative-earth arrangement. Mid-point and positive earthing arrangements may alternatively be used.

Note 7: Incoming/outgoing mains and DC connections are omitted for simplicity. Mains connections are discussed in Sections 9.3 and 9.4, and DC connections are discussed in Section 9.10.

Note 8: The charger, controller and inverter will require output protection, which may be a combination of electronic protection and dedicated protective devices. For simplicity, these are not shown.

Figure 9.2   Example of EESS ELV battery and PCE in same enclosure

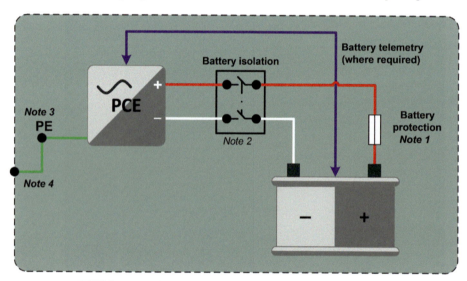

Notes to Figure 9.2

Note 1: Where the battery and PCE are located in the same sealed enclosure, and maintenance of the storage battery on site is not envisaged, an isolator for the storage battery is not always necessary. However, where required, see Section 9.2.1(a), to be located physically and electrically as close as possible to the battery. Must be rated for DC protection, and be capable of breaking at least the maximum short-circuit current of the battery. Whilst a fuse is shown, an MCB or MCCB, rated for DC protection, is an equally acceptable alternative.

Note 2: Located physically and electrically as close as possible to the battery, after the battery protection fuse (see Note 1). Double-pole isolation is recommended for battery voltages above 48 V DC and is essential for LV batteries (above 120 V DC). An on-load isolator should be specified where the battery may be replaced/maintained with the system still operational. Where the system is to be isolated before battery maintenance/replacement, a plug and socket arrangement may be suitable.

Note 3: LV protective conductor.

Note 4: Connections for earthing of conductive enclosures.

Note 5: For simplicity, incoming/outgoing mains and DC connections are omitted. Mains connections are discussed in Sections 9.3 and 9.4, and DC connections are discussed in Section 9.10.

Note 6: In this figure, the batteries are shown in an unearthed arrangement. Battery earthing may not be required provided there is adequate means of protection against electric shock, for example, with ELV batteries.

Section 9 – Design of an EESS

9.2.2 Battery and BMS protection

Due to the high short-circuit currents available, and the nature of a DC fault, it is advisable for connections between the PCE or BMS and battery to be suitably protected against damage, such as impact. This is particularly important where the PCE or BMS and battery have separate enclosures. Since short-circuit currents from ELV batteries are typically high (of the order of kA), and at ELV the impedance of the fault may prevent the protective devices from operating (with LV, the fault has negligible impedance compared with the other circuit impedances), it is recommended that, where the battery voltage is ELV, suitable earthed metal conduit, trunking, or armoured cables are used.

Protective and switching devices for DC circuits need to be suitable for the purpose. Some DC rated equipment is not rated for bi-directional current flow, which is essential for many DC circuits in energy storage systems. Many protective and switching devices used for AC circuits are not suitable for DC circuits. Further guidance is available in the IET Technical Briefing *Practical considerations for DC installations*.

In the general case, both battery and charger will contribute to faults in the circuitry and wiring between the charger, battery and inverter. Ideally, the charger will have electronic overcurrent protection (backed up by a suitable fuse or circuit breaker).

The system should have a suitable high-breaking capacity protective device, rated for the maximum prospective fault current in accordance with the manufacturer's instructions, mounted close to the terminals of the battery assembly, as well as an isolator. Double-pole isolation is recommended for battery voltages above 48 V DC and is essential for LV batteries (above 120 V DC). The isolator may be formed to include a suitable DC rated circuit breaker, which should be arranged to disconnect all poles.

Note: Where the battery and PCE are located in the same sealed enclosure, and maintenance of the storage battery on site is not envisaged, an isolator for the storage battery is not always necessary.

9.2.3 DC Earthing

9.2.3.1 General

Protective and functional earthing should be provided for all connections between battery and PCE enclosures, where these are separate. Enclosures, conduit and trunking may be used as protective conductors in accordance with Chapter 54 of BS 7671.

Note: The designer may opt to use an IT system with insulation monitoring, where this meets the requirements of BS 7671 and BS EN IEC 624855-2, but this is not recommended for EESS in dwellings. Earthing may not be required where the battery, BMS and PCE are located in the same enclosure, such as in the example provided in Figure 9.2.

Protective earthing in accordance with BS 7671 is essential for batteries operating at LV. This must be connected to a means of earthing in accordance with BS 7671 and BS 7430, and it is recommended that a bond to the LV protective conductor is provided where simple separation is included in the PCE. Where the PCE does not contain simple separation, the designer of a system being assembled from discrete components should check that the battery arrangement complies with BS 7671 and has an appropriate means of protection against electric shock and fault current.

The requirements of PELV or FELV in BS 7671 should be maintained for batteries operating at ELV. Unearthed ELV batteries should meet the BS 7671 requirements for SELV.

It is important to note that, if the system is being assembled from discrete components, it may not be possible to solidly bond the DC functional earth to the LV protective conductor unless the PCE contains simple separation.

Note: 'Simple separation' is defined in BS 7671 as "separation between circuits or between a circuit and earth by means of basic insulation". The requirements of basic insulation for simple separation are given in Section 5.3.3 of BS EN 60664-1.

Section 9 – Design of an EESS

Selection of the method of DC earthing should consider the risk of causing corrosion to structural steelwork in contact with the general mass of Earth, along with degradation of extraneous-conductive-parts.

The designer may also choose solid earthing, high resistance earthing, diode earthing or capacitive earthing, to help reduce the impact on connected networks and systems from DC line to Earth and line to line faults.

9.2.3.2 Mid-point earthing

An example of mid-point earthing is included in Figure 9.1(a). Mid-point earthing can be used for:

(a) minimizing galvanic corrosion effects of earthed/bonded metalwork, such as buried metallic services;
(b) reducing the voltage to earth of the DC system; and
(c) earthing of certain non-isolated converter types.

Mid-point earthing requires overcurrent protective devices in both L+ and L- poles for both automatic disconnection and protection against fault current.

9.2.3.3 Negative earthing

An example of negative earthing is included in Figure 9.1(b). Negative earthing is widely used for lead-acid battery installations in many applications and voltage ranges.

In 2-wire earthed DC systems, it helps reduce the susceptibility to corrosion of battery terminals and earthing terminals/conductors that may occur in positive earthed DC systems.

In large (high current/high power) systems, stray leakage currents from any 2-wire DC circuit may cause corrosion of buried metalwork and services, whether or not they form part of the electrical installation, and galvanic protection may therefore be required.

9.2.3.4 Positive earthing

An example of positive earthing is included in Figure 9.1(c). There may be occasions where a manufacturer opts for positive earthing for reasons of monitoring and/or electronic control.

Under certain conditions, positive earthing may increase the susceptibility for corrosion of battery terminals and earthing terminals/conductors.

In large (high current/high power) systems, stray leakage currents from any 2-wire DC circuit may cause corrosion of buried metalwork and services, whether or not they form part of the electrical installation, and galvanic protection may therefore be required.

9.2.4 Cable identification for DC cabling

9.2.4.1 Wiring in the electrical installation

The identification of conductors of cables by colours and alphanumeric marking is specified in Table 51 of BS 7671, which aligns with the harmonized standard BS EN 60445. The single colour green must not be used to identify conductors in AC or DC power cabling installed on site.

To avoid confusion within an installation, it is recommended to separately identify DC conductors from AC conductors by the use of additional alphanumeric marking on the conductors. Cables or containment can likewise be marked or labelled, clearly stating 'DC' on the marking for DC cables. It is recommended that wiring systems consisting of singles in containment do not mix AC and DC LV cables in the same containment. ELV cables would normally be separately contained from LV cables.

Section 9 – Design of an EESS

9.2.4.2 Wiring in pre-manufactured assemblies

Products standards and BS EN 60445 permit slight variations from BS 7671 wiring colours and, particularly for ELV DC wiring, a variety of options may be selected by a manufacturer of pre-manufactured assemblies to suitable standards.

Table 9.2   DC wiring colours complying with BS 7671 and BS EN 60445

DC conductor function	Conductor identification	
	Colour	Alphanumeric
Protective earth	Green-and-yellow (GNYE)	PE
Functional earth	Pink (PK)	FE
Positive of two-wire circuit	Red (RD)	L+
Negative of two-wire circuit	White (WH)	L-
Positive of two-wire negative-earth circuit	Red (RD)	L+
Negative of two-wire negative earth circuit	Blue (BU)	M
Positive of two-wire positive earth circuit	Blue (BU)	M
Negative of two-wire positive earth circuit	White (WH)	L-
Positive of three-wire circuit	Red (RD)	L+
Mid-point or earthed mid-point of three-wire circuit	Blue (BU)	M
Negative of three-wire circuit	White (WH)	L-

BS EN 60445 and BS 7671 reserve the following colours for specific purposes:

(a) the colour-combination green and yellow for protective earthing conductors; and
(b) the colour blue for earthed current carrying conductors (neutral, earthed mid-point and earthed DC current carrying conductors).

9.3   The EESS operating in parallel with the grid (connected direct-feeding and connected reverse-feeding modes)

9.3.1   General

The principal issues with this mode of operation are:

(a) the EESS can contribute to the available fault current; and
(b) there are operational scenarios in which it is hazardous for the EESS to continue to supply power in parallel with the grid.

A non-exhaustive list of aspects that need to be considered in the distribution board specification are:

(a) the maximum short-circuit level (parallel energy sources plus grid-connected);
(b) the minimum short-circuit level (typically islanded mode) to achieve related disconnection times, including thermal constraints for cables/busbars and shock protection; and
(c) circuits potentially supplied by both sides and/or components becoming reverse-connected. Protective devices that are suitable for bi-directional overcurrent detection and current flow shall be selected.

Some protective devices are polarized and are intended to supply loads in one direction of current flow only. Devices such as a polarized circuit-breaker or polarized residual current protective device will have clearly marked line and load terminals.

Section 9 – Design of an EESS

9.3.2 Protection against electric shock

Note: Chapter 41 of BS 7671 is the primary reference for automatic disconnection for protection against electric shock.

The following principles should be adopted:

(a) the EESS should be integrated into the installation so that it does not impact the safety of the installation in parallel modes of operation.

(b) an RCD on the AC side at, or close to, the EESS inverter is recommended in TT systems, to ensure automatic disconnection.

(c) the EESS should not provide additional connections between neutral and earth conductors; this is prohibited by BS 7671 to prevent nuisance-tripping of RCDs and protective conductors from carrying load currents. Where the EESS is intended to operate in island mode, the neutral to the grid supply should be broken immediately before the neutral to earth bond is made at substantially the same time. See Section 9.4.7.

(d) where RCDs are present in the installation, they should be compatible with the EESS inverter (see Section 9.12).

(e) the EESS inverter should not be connected downstream of RCDs providing additional protection to multiple circuits. For example, an EESS converter should not be connected to an MCB downstream of an RCD that is shared by other protective devices in a split-load consumer unit.

9.3.3 Fault current protection for the installation

Note: Chapter 43 of BS 7671 is the primary reference for fault current protection, and Chapter 42 of BS 7671 deals with protection against thermal effects.

Considerations for fault current protection in parallel operation will depend on the power delivery capability of the EESS inverter as well as the energy available from storage. In the majority of smaller systems, a large part of the fault current would normally be provided by the grid. A larger EESS may provide a significant contribution over the standing prospective short circuit current.

Inverters or converters associated with PV, fuel cell and energy storage, do not behave the same as synchronous or induction machines. They do not have a rotating mass component; therefore, they do not develop inertia and have a much faster decaying envelope for fault currents. An accepted engineering practice to approximate the short-circuit contribution of inverters or converters associated with PV, fuel cell and energy storage is 1.2 times its rated current added to the fault current normally provided by the grid. Where more than one invertor can supply fault current simultaneously, it is the sum of 1.2 times their rated currents.

Under conditions where the EESS contributes significantly to the fault current, the designer (or manufacturer) should determine whether to temporarily disconnect the EESS (similar to 'bypass' mode on a UPS, where if a fault is detected, the inverter is disconnected but the mains remains connected) to permit the grid supply to clear the fault.

9.3.4 Overload protection

9.3.4.1 Overload protection for circuits in the installation

Overload protection for final circuits in the installation is provided in accordance with BS 7671. The design should ensure coordination of the relevant overcurrent protective device with the current carrying capacity of the cable.

Section 9 – Design of an EESS

9.3.4.2 Overload protection for the circuit to which the EESS is to be connected

When an EESS is operating in parallel with the grid, there will be at least one part of the installation in which the conductors share current from both the EESS and grid, and this part of the installation is not protected against overcurrent by the usual method described in 9.3.4.1. Outside simple installations in dwellings, larger systems use different protection methodologies and unprotected sections of cable or busbar should be avoided.

Installing EESS onto existing circuits changes the nature of the circuit and the current carrying capacity of the existing conductors may not be sufficient (see BS 7671 Regulations 551.7.1 and 551.7.2). It is unlikely to be within the capability of ordinary persons and general DIY enthusiasts to carry out the necessary design work.

Parts of the installation supplied by two sources and not protected by a single overcurrent protective device should be arranged to ensure they comply with BS 7671, in particular Section 551.7 (see Figure 9.3).

The recommendations of the manufacturer of the consumer unit or distribution board should be followed when considering the rating of busbars, and the positioning of protective devices based on loading and duty cycle, to prevent overheating of the assembly and/or busbar.

Where the EESS is to be connected via a switchgear or controlgear assembly, for example, a consumer unit or distribution board, the assembly rating I_{nA} should be selected so that:

$$I_{nA} \geq I_n + I_{g(s)} \text{ where:}$$

I_{nA}	is the rated current of the assembly
I_n	is the rated current of the overcurrent protective device supplying the incoming isolator of the assembly
$I_{g(s)}$	is the rated output current of all forms of generation connected to the load side of the assembly.

This is illustrated in Figure 9.3.

The rating of a switchgear or controlgear assembly is particularly important where AC coupled systems operate in installations where there are other forms of generation. Figure 9.4 shows an example of AC coupled EESS connected to a consumer unit that also has solar PV connected to it, and also an example of a main distribution, serving two final circuit distribution boards each with an EESS connected.

Figure 9.3 Examples of parts of installations whose rated current may be impacted by the introduction of EESS or other embedded generation

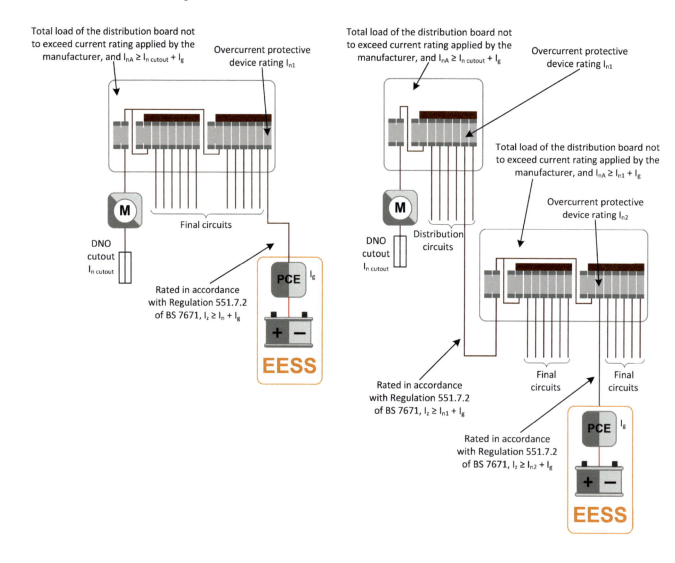

Section 9 – Design of an EESS

Figure 9.4 Examples of rating of a distribution board having more than one generation source is connected through it

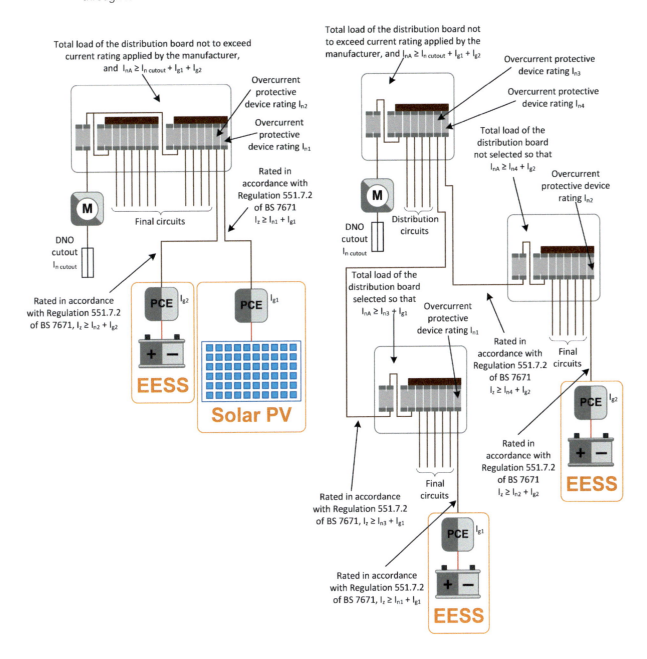

9.3.4.3 Single-phase EESS connected into three-phase systems

If a single-phase inverter is connected on one phase only, is exporting, and that phase is lightly loaded, the neutral current in parts of the installation, or the incoming grid supply, may far exceed the line currents. It is therefore preferable to install three-phase inverters for three installations. Where single-phase inverters are used to export on one or two phases, load currents in the installation should be monitored, and the export of the inverters curtailed to prevent excessive neutral currents.

9.4 The EESS as an alternative source of supply to the grid (island mode)

9.4.1 General

The principal issues with island mode operation are:

(a) the EESS is likely to have considerably lower prospective fault current than the grid supply to operate protective devices;

(b) the EESS cannot rely on the means of earthing from the grid, as this is removed, for example, during maintenance or faults in the grid; and

(c) smaller EESS may not be capable of providing long-term supplies to high power loads, and load shedding may therefore be required.

In some electrical installations, more than one means of earthing may be utilized, often to meet the requirements of Part 7 of BS 7671 for certain parts of the installation. For example, an installation with TN-C-S supply may contain circuits supplying outbuildings, electric vehicle charging points, hot tub, or caravan connection outlets, which have TT earthing. The supply earthing arrangements for island mode operation must be selected carefully to ensure that such circuits for special locations that are not isolated during island mode operation continue to meet their safety requirements.

Note: Section 9.4 is also applicable to grid-independent systems where the primary generation source is offline, although grid-independent systems can be implemented with one means of earthing.

9.4.2 Systems that switch between grid-connected and island mode operation

In designing a system that switches between on-grid (grid-connected) and island mode operation, the following factors need careful consideration:

(a) the general earthing arrangements on- and off-grid. These may not be the same: for example, a system that is TT on-grid may be TN-S when switched to island mode. See Section 9.4.4.

(b) the configuration, rating and functionality of the device that isolates the grid in order that island mode operation can commence. See Section 9.4.9.

(c) considerations for a neutral-earth (N-E) bond during island mode operation. Switching in an N-E bond during island mode operation is likely to form part of the switching arrangements described above. See Section 9.4.6.

(d) the means of earthing during island mode operation. In many cases, an earth electrode will need to be installed to meet these requirements. See Section 9.4.10.

(e) that disconnection times, for all circuits to be maintained in island mode operation, can be met without the grid supply in island mode. See Section 9.4.11.

9.4.3 General considerations for earthing arrangements

Regulation 551.4.3.2.1 of BS 7671 requires that the consumer's own earthing arrangements must be connected when the system switches to island mode operation. The consumer's means of earthing may remain connected in connected modes of operation, in installations' TN-C-S and TN-S grid supplies.

> **551.4.3.2.1** Protection by automatic disconnection of supply shall not rely upon the connection to the earthed point of the system for distribution of electricity to the public when the generator is operating as a switched alternative to a TN system. A suitable means of earthing shall be provided.

As consumers are not permitted to combine neutral and earth functions in conductors in installations that would normally be connected to the supply from the distribution network operator (DNO), the design decision for the majority of EESS operating at low voltage is whether the means of earthing in island mode should be TN-S or TT.

Section 9 – Design of an EESS

Existing TN-C-S and TN-S installations may not have existing consumer earth electrodes, and these will therefore need to be installed along with the EESS – see Section 9.4.10.

TT installations will have existing consumer earth electrodes, but their suitability for earthing of the EESS in island mode should be ascertained prior to installation of the EESS, and upgraded where necessary.

9.4.4 TN-S island-mode arrangement

TN-S generally provides a lower earth fault loop impedance (Z_s) than TT in island mode, and is therefore preferred. The actual means of earthing arrangement selected will depend on one or more of the following factors:

(a) the design of the system itself, for example, whether it contains isolating transformer outputs, and the neutral switching options available;

(b) the 'on-grid' means of earthing and the availability of earth electrodes; and

(c) the energy available to operate protective devices (short circuit currents and loop impedances) when operating in island mode.

Operating as TN-S in island mode ensures that loop impedances for earth faults are minimized. Where the supply arrangements are TT, operating as TN-S in island mode offers improved protection against earth faults, and therefore this is recommended. Figure 9.5 provides a simplified illustration of the earthing arrangements in island mode, for TN-C-S, TN-S and TT connected-mode earthing arrangements.

9.4.5 IT island mode arrangement

With the exception of very small installations, an IT system is rarely feasible for island mode operation, for a variety of reasons, including the following:

(a) commercially available appliances and equipment are often Class I. A means of detecting a first fault would be imperative.

(b) many items of equipment in use in modern installations have interconnectivity options that may import Earth potential. For example, a computer connected to a device via a screened data cable that is connected at the other end to a PELV system would provide an earth connection.

9.4.6 N-E bond arrangements

In most situations, switching into island mode will mean that the grid N-E bond will be lost and a new N-E bond will need to be provided for the duration of island mode.

It is important to ensure that there is no more than one connection between neutral and earth in the system at any one time, to prevent unwanted operation of RCDs, and to reduce risks associated with surges and induced voltages. Regulation 551.6.2 should also be considered.

> **551.6.2** For a TN-S system where the neutral is not isolated, any RCD shall be positioned to avoid incorrect operation due to the existence of any parallel neutral-earth path.
>
> **Note:** It may be desirable in a TN system to disconnect the neutral of the installation from the neutral or PEN of the system for distribution of electricity to the public to avoid disturbances such as induced voltage surges caused by lightning.

Note: Where the DNO supply to an installation is provided at LV, disconnection of the supply neutral by the island mode isolator will be required to comply with ENA Engineering Recommendations G98 and G99.

Consequently, in TN and TT systems, it is necessary to provide suitable switching arrangements to ensure that:

(a) all live conductors, including the neutral, of circuits to be powered in island mode are disconnected from the neutral or PEN of the DNO supply; and

(b) a single N-E bond exists in either island mode or on-grid mode of operation.

Section 9 – Design of an EESS

A switch that provides an N-E bond during island-mode operation should be interlocked and operate as described in Section 9.4.7.

9.4.7 Switching arrangements to establish TN-S in island-mode operation

Transfer between connected mode and island mode, and vice versa, can be achieved by operating switching devices for islanding. This can be either directly controlled (manually or remotely) or automatically controlled.

Switching arrangements should incorporate the following:

(a) the island mode isolator should always break the neutral as well as other live conductors to prevent other consumers' neutral currents from being shared with the installation's earthing system.

(b) the switching arrangements that provide the N-E bond during island mode operation should be interlocked with the island mode isolator in accordance with Regulation 537.1.5 of BS 7671. When moving to island mode, the contact should be closed immediately after the live conductor contacts of the island mode isolator are opened. When moving to connected mode, the contact should be opened immediately before the live conductor contacts of the island mode isolator are closed.

(c) in polyphase systems, the neutral contact of the island mode isolator should not disconnect before those of the line conductors, and should not reconnect after those of the line conductors, in accordance with Regulation 431.3 of BS 7671.

(d) the rating of the switching devices should be as per Section 9.4.9.

Arrangements for isolation and earthing in island mode are illustrated as follows:

(a) Figure 9.5 is a simplified illustration showing earthing and switchover arrangements in connected mode and island mode, to achieve a TN-S island mode arrangement.

(b) Figure 9.6 provides an example of the timing sequence required for the island mode isolator and N-E bond relay.

(c) Figure 9.7 shows an arrangement where the entire installation is to be powered in island mode. The island mode isolator essentially operates at the origin.

(d) Figure 9.8 and Figure 9.9 illustrate arrangements where only selected loads are maintained in island mode. The island mode isolator operates to isolate the part of the installation containing the EESS, and maintained loads, from the remainder of the installation and the grid.

These arrangements align with Clause 7.2.6 of BS 7430. They are, however, not the only options available to the designer. For simplicity, only single-phase small-scale systems are shown to illustrate the principles.

551.4.3.2.1 Protection by automatic disconnection of supply shall not rely upon the connection to the earthed point of the system for distribution of electricity to the public when the generator is operating as a switched alternative to a TN system. A suitable means of earthing shall be provided.

551.6.2 For a TN-S system where the neutral is not isolated, any RCD shall be positioned to avoid incorrect operation due to the existence of any parallel neutral-earth path.

Note: It may be desirable in a TN system to disconnect the neutral of the installation from the neutral or PEN of the system for distribution of electricity to the public to avoid disturbances such as induced voltage surges caused by lightning.

Note: Where the DNO supply to an installation is provided at LV, disconnection of the supply neutral by the island mode isolator will be required to comply with ENA Engineering Recommendations G98 and G99.

Section 9 – Design of an EESS

Figure 9.5   Simplified illustration showing earthing and switch-over arrangements in connected mode and island mode, for TN-S island-mode arrangement

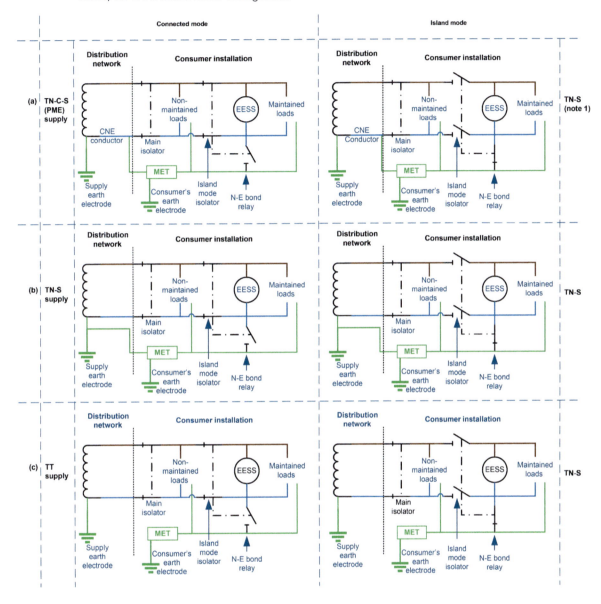

Note 1: For all connected mode supply earthing arrangements, it is recommended to switch to a TN-S earthing arrangement in island mode. Where PME conditions apply in connected mode, these conditions will remain in island mode, because for TN-C-S and TN-S supplies, the installation remains connected to the distributor's means of earthing.

Note 2: A TT earthing arrangement in connected mode becomes a TN-S arrangement in island mode.

Note 3: All arrangements illustrate that there is only a single N-E bond in place for maintained circuits in island mode, and the island mode isolator prevents diverted and circulating neutral currents.

Note 4: The island mode isolator should always break the neutral as well as other live conductors.

Note 5: The N-E bond relay is interlocked with the island mode isolator. When moving to island mode, it is closed at the same time as, or immediately before, the live conductor contacts of the island mode isolator are opened. When moving to connected mode, the N-E bond relay is opened at the same time as, or immediately before, the live conductor contacts of the island mode isolator.

Note 6: In installations with TN-C-S (PME) supplies, the consumer's earth electrode may not be present prior to the installation of EESS, and a new electrode will be installed. See Section 9.4.10.

Section 9 – Design of an EESS

Figure 9.6   Example of timing arrangements for the island mode isolator and N-E bond relay

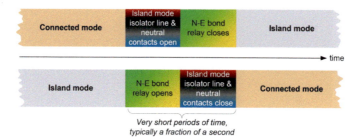

Figure 9.7   Arrangement, island mode isolation and earthing, where entire installation is supplied in island mode

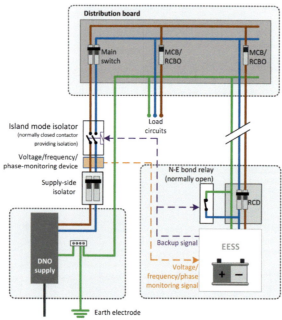

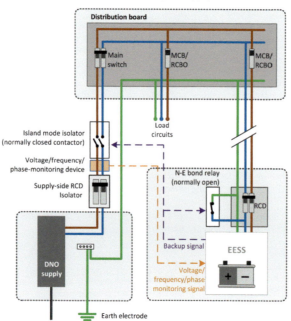

Section 9 – Design of an EESS

Note 1: The terms 'normally open' and 'normally closed' refer to the state 'on-grid', but contacts are shown in the island mode position.

Note 2: Section 9.4.10 outlines the criteria for earth electrodes, including assessing whether an existing consumer's earth electrode for a TT system is suitable for EESS island mode operation, or should be renewed when the EESS is installed.

Note 3: In this arrangement, the designer and installer must ensure that the consumer knows the limitations of the storage for powering large loads when operating in island mode.

Note 4: The RCD at the output of the EESS in TN systems may not be required, where Regulation 419.2 of BS 7671 applies.

Figure 9.8   Arrangement, island mode isolation and earthing, where only maintained loads are supplied in island mode

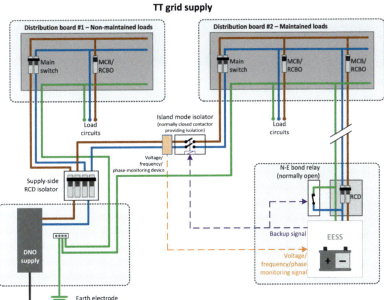

Note 1: The terms 'normally open' and 'normally closed' refer to the state 'on-grid', but contacts are shown in the island mode position.

Note 2: Section 9.4.10 outlines the criteria for earth electrodes, including assessing whether an existing consumer's earth electrode for a TT system is suitable for EESS in island mode operation, or should be renewed when the EESS is installed.

Note 3: The RCD at the output of the EESS in TN systems may not be required, where Regulation 419.2 of BS 7671 applies.

Section 9 – Design of an EESS

Figure 9.9 Alternative arrangement, island mode isolation and earthing, where only maintained loads are supplied in island mode

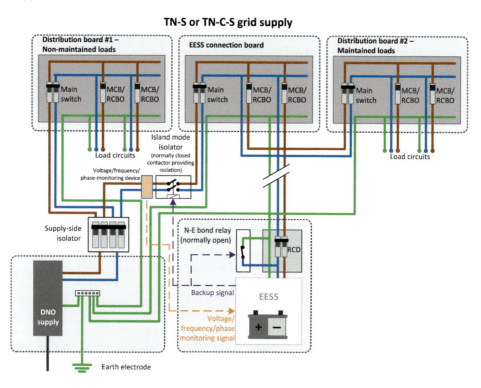

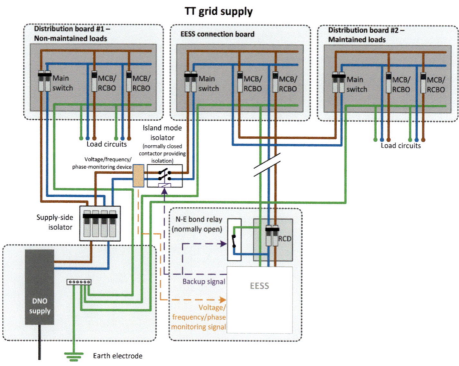

Note 1: The terms 'normally open' and 'normally closed' refer to the state 'on-grid', but contacts are shown in the island mode position.

Note 2: Section 9.4.10 outlines the criteria for earth electrodes, including assessing whether an existing consumer's earth electrode for a TT system is suitable for EESS in island mode operation, or should be renewed when the EESS is installed.

Note 3: The RCD at the output of the EESS in TN systems may not be required, where Regulation 419.2 of BS 7671 applies.

9.4.8 Location of N-E bond relay, where an RCD is provided at the EESS output

Where an RCD is used at the output of an EESS (as shown in Figures 9.7, 9.8 and 9.9; and as described in Section 9.4.7), the location of the N-E bond needs to be considered. For this RCD to operate and protect the installation downstream of the EESS in island mode, the N-E bond needs to be between the RCD and the EESS.

9.4.9 Rating the island mode isolator and N-E bond relay

The island mode isolator should be rated to take into account the following:

(a) the line and neutral contacts of the island mode isolator should be rated for an appropriate design current in line with Section 9.3.4.2, considering parallel operation. The rating will depend on the point in the installation at which the island mode isolator is installed, for example:
- **(i)** for an island mode isolator installed at the origin (for example, Figure 9.7), the rating will depend on the maximum demand of the installation (the sum of the loads, considering diversity, and the total generation within the installation); and
- **(ii)** for an island mode isolator that isolates only part of the installation (for example, Figures 9.8 and 9.9), the rating will depend on the maximum demand of the distribution board(s) supplied by the island mode isolator (the sum of the maintained loads, considering diversity, and the total generation remaining connected in island mode).

(b) the line and neutral contacts of the island mode isolator should be capable of making or breaking onto a fault equivalent to the sum of the prospective fault currents of all sources of energy operating in parallel. Typically, this will be the largest sum of the prospective currents from all sources capable of simultaneously operating in parallel.

(c) it is recommended that the neutral contacts are not de-rated for polyphase systems. In systems where increased sized neutrals are incorporated, this should be considered for the neutral contact. In polyphase systems, the neutral contact of the island mode isolator should not disconnect before those of the line conductors, and should not reconnect before those of the line conductors, in accordance with Regulation 431.3 of BS 7671.

The island mode isolator and N-E bond relay should be selected so that their anticipated reliability aligns with the recommended inspection and test interval for the installation.

The N-E bond relay contact should be rated as follows:

(a) where the grid supply is a TN system, the N-E bond relay contact should be rated for the sum of all generation in the installation that may operate in island mode;
(b) where the grid supply is TT, the N-E bond relay should be rated as per the line contacts of the island mode isolator; and
(c) the N-E bond relay contact should be capable of making or breaking onto a fault equivalent to the sum of the prospective fault currents of all sources of energy operating in parallel. Typically, this will be the largest sum of the prospective currents from all sources capable of simultaneously operating in parallel.

In most electrical installations, currents do flow in the protective conductor under normal conditions. Protective conductor currents may be seen to approach a few amps in a dwelling, and tens of amps in installations with higher demand.

Protective conductor currents that flow in normal conditions may result from:

(a) the correct operation of filters in electrical machinery, appliances and electronic equipment to prevent conducted electromagnetic disturbances.
(b) screens of cables discharging electromagnetic energy to earth. This increases with the length and quantity of the screened cables connected to the earthing system, and with the field strength of the electromagnetic sources to which the cables are subjected.

(c) small currents flowing through insulation in cables, motors, heating elements, etc., or due to capacitive effects. In smaller installations such as dwellings, current flow in the protective conductor due to 'leakage' via insulation and capacitive effects is relatively small, although this increases with the size of the installation.

Under certain circumstances, particularly in TN-C-S (PME) connected systems with a broken neutral in the grid network, load currents may be seen to flow through protective conductors. In earth faults, fault currents necessarily flow through protective conductors.

The functions of the island mode isolator and N-E bond relay in a simple installation could be facilitated with a single multi-pole switch. Such a device should be selected to meet the requirements of Regulations 431.3, 537.1.5 and 543.3.3.101 of BS 7671.

> **419.2** For installations with power electronic convertors with nominal voltage U_0 greater than 50 V AC or 120 V DC and where automatic disconnection is not feasible, the output voltage of the source shall be reduced to 50 V AC or 120 V DC or less in the event of a fault between a live conductor and the protective conductor or Earth in a time as given in Regulation 411.3.2.2, 411.3.2.3 or 411.3.2.4, as appropriate (see BS EN 62477-1).
>
> The power electronic convertor shall be one for which the manufacturer gives adequate methods for the initial verification and periodic inspection and testing of the installation.

9.4.10 The means of earthing (consumer's earth electrode)

Regulation 551.4.3.2.1 of BS 7671 requires that the consumer has their own earth electrode, and the supplier's means of earthing cannot be relied upon in island mode:

> **551.4.3.2.1** Protection by automatic disconnection of supply shall not rely upon the connection to the earthed point of the system for distribution of electricity to the public when the generator is operating as a switched alternative to a TN system. A suitable means of earthing shall be provided.

For TN systems, EESS operating in island mode cannot rely on the supplier's means of earthing.

For TT grid-supply arrangements, an existing consumer's earth electrode with high earth electrode resistance (Z_{EE}) may be unsuitable for an EESS in island mode operation.

Consumer's earth electrodes should in general comply with BS 7671 and BS 7430. Earth electrodes dedicated to lightning protection should not ordinarily be used as the means of earthing for the EESS. Foundation earthing conductors may be suitable for use by the EESS if there is no possibility of their galvanic degradation.

Whilst the supplier's means of earthing cannot be relied upon during island mode operation, it should remain connected for the following reasons:

(a) complete separation of the island mode earthing system from the DNO's earthing system is not possible, especially in small curtilage properties;
(b) there are advantages in remaining connected to a global earthing system where one is provided (see BS EN 50522 for further details); and
(c) extraneous-conductive-parts, particularly metallic service pipes, may be common to other installations, and any disconnection would be ineffective.

The earth electrode resistance (Z_{EE}) should be as low as practicable, and should not exceed 200 Ω. Earth electrode resistances above 200 Ω are likely to be unstable and are therefore unsuitable for EESS. The type and embedded depth of an earth electrode should be such that soil drying and freezing will not increase its resistance above the required value.

Section 9 – Design of an EESS

The flowchart in Figure 9.10 can be used to help determine the requirements for earth.

Figure 9.10 Island mode earth electrode selection

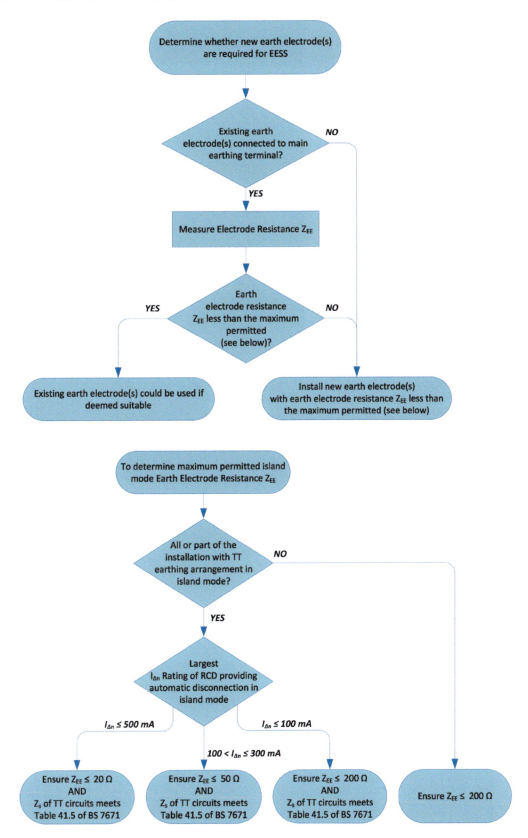

Section 9 – Design of an EESS

In circuits with a TT earthing arrangement in island mode (for example, where required for installations with EV charging), and in which RCDs provide automatic disconnection, the earth loop impedance of foreseeable earth faults should not exceed the values in Table 41.5 of BS 7671 in relation to the largest residual current rating of RCDs providing automatic disconnection (see Table 9.3).

Figure 9.11 provides an example of an installation that has both TN and TT arrangements maintained in island mode. In some instances, where the grid supply has a TN-C-S or TN-S earthing arrangement, it may not be possible for a vehicle to grid system to remain connected in island mode, unless a separate TT earthing arrangement for the vehicle can be safely provided.

Table 9.3 Recommended maximum earth fault loop impedances for final circuits in TT systems operating in island mode, based on the residual current rating of the RCD providing automatic disconnection

Residual current rating of RCD providing automatic disconnection ($I_{\Delta n}$) (mA)	Maximum Z_s for the final circuits protected (Ω) See Table 41.5 of BS 7671
30	1,667
100	500
300	167
500	100

Note: The resistance of the installation earth electrode(s) should be as low as practicable. Earth electrode resistances exceeding 200 Ω may not be stable.

In the case of a system with EV charging, where there are separate earth electrodes provided for the EESS and the EV charger:

(a) the earth electrode for the EV charging equipment must be separated below ground by a distance of at least 2.5 m (or at least 2.0 m for on-street EV charging equipment installations) from buried conductive parts connected to the PME earthing system. Some DNOs may require a greater distance.

(b) the IET *Code of Practice for Electric Vehicle Charging Equipment Installation* advises on the separation of earthing systems below ground, and the need to avoid simultaneous contact between exposed-conductive-parts connected to different earthing systems.

Safety and earthing requirements for vehicle to grid systems should comply with the IET *Code of Practice for Electric Vehicle Charging Equipment*, 4th Edition.

Section 9 – Design of an EESS

Figure 9.11   Example consideration of island mode earth fault loop paths

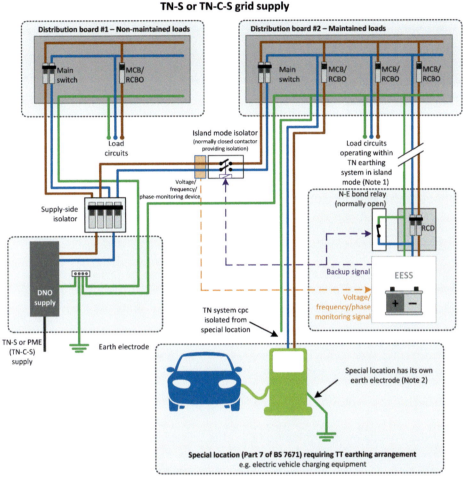

Note 1: Earth fault loop impedances required to meet Table 41.5 of BS 7671. The island mode earth fault return path for the system in island mode will return through the protective earthing system, via the main earthing terminal, and through the N-E bond of the EESS.

Note 2: Earth fault loop impedance of TT circuits required to meet Table 41.5 of BS 7671. The island mode earth fault return path for the system in island mode will return through the TT earth electrode, to the EESS earth electrode, via the main earthing terminal, and through the N-E bond of the EESS.

9.4.11 Automatic disconnection of supply for protection against electric shock

The designer should assess whether the disconnection times, for all circuits to be maintained in island mode operation, can be met without the grid supply. Since there is likely to be a lower prospective fault current in island mode, overcurrent protective devices may not deliver an adequate disconnection time.

RCDs may be used to provide automatic disconnection of supply (see Regulation 411.4.5); however, overcurrent protective devices are usually required in addition. In dwellings, newer installations are more likely to incorporate RCDs, generally with a residual operating current of 30 mA, for additional protection. Installations in older dwellings may require RCDs to be retrofitted. All existing RCDs should be checked to ensure that they can operate with any DC components that may be present; see Section 9.12.

Regulation 419.2 permits PCE to provide protection where it is not feasible to use a protective device.

Section 9 – Design of an EESS

> **419.2** For installations with power electronic convertors with nominal voltage U_0 greater than 50 V AC or 120 V DC and where automatic disconnection is not feasible, the output voltage of the source shall be reduced to 50 V AC or 120 V DC or less in the event of a fault between a live conductor and the protective conductor or Earth in a time as given in Regulation 411.3.2.2, 411.3.2.3 or 411.3.2.4, as appropriate (see BS EN 62477-1).
>
> The power electronic convertor shall be one for which the manufacturer gives adequate methods for the initial verification and periodic inspection and testing of the installation.

Where the entire installation is designed to operate with a TT supply arrangement in island mode, for example, where this is necessary to facilitate vehicle to grid, an RCD is required at the EESS inverter output to protect against earth faults that may occur between the EESS and the RCD.

9.4.12 Fault current protection

Note: Chapter 43 of BS 7671 is the primary reference for protection against overcurrent, and Chapter 42 of BS 7671 deals with thermal effects.

The EESS requires back-up protection because it may not be able to provide adequate fault current to operate final circuit overload protection existing in the installation. Electronic protection should be arranged to prevent automatically re-closing onto faults.

Where cables and their protective devices are selected in accordance with the requirements of BS 7671, additional fault current protection is not generally required. The criteria for thermal effects is related to the cable heating under fault conditions, which is addressed in BS 7671 using the adiabatic criterion:

$$k^2 S^2 \geq I^2 t$$

The designer of the installation incorporating EESS that is intended to operate in island-mode should consider that the EESS will have a lower prospective short circuit current available in this mode, and protective devices for fault protection may not operate before the stored energy is expended.

Electronic means of switching off the EESS inverter in the event of a fault may be employed, in line with Regulation 419.2 of BS 7671, but these are not selective and power to the whole installation may be removed because of a fault on a single circuit. Whilst this may be acceptable in some installations, there are occasions where the designer of the EESS installation must consider the consequences of removing power from the whole installation because of a fault on a single circuit, and how power might be reinstated (for example, 'hiccup' restart attempts).

9.5   Selectivity with supply-side protective devices

Whilst selectivity between protective devices typically ensures that a device within the installation will operate first, there are occasions where the fault current is of such a magnitude that a supply-side protective device (such as the supplier's cut-out, or for private distribution networks, a feeder pillar fuse) may operate before protective devices within the installation.

9.5.1   Installations up to 100 A per phase and total EESS not exceeding 6 kVA

Selectivity issues are more likely to occur with higher prospective fault currents and lower external earth fault loop impedances, such as installations close to a distribution transformer.

Where the protective devices within the installation are MCBs to the BS EN 60898 series, RCBOs to the BS EN 61009 series, or cartridge fuses, a fault that operates the supplier's cut-out first will be of such a magnitude that the supplier's cut-out operates in a relatively negligible time. Once the system switches to island mode, the available fault current is far lower. Any remaining fault will be detected either by the PCE's internal overcurrent protection or the RCD at the PCE output, and the impact of switching over is likely to be minimal in terms of clearing the fault within safe parameters.

In an installation with re-wireable fuses, for example, to BS 3036, the supplier's cut-out may operate first with lower fault currents, some perhaps persisting for a much longer relative time before disconnection.

EESS should not be retrofitted to installations where final circuits and distribution circuits protected by re-wireable fuses are powered by the EESS during island mode operation.

9.5.2 Installations over 100 A per phase and/or total EESS greater than 6 kVA

The designer should consider either:

(a) the coordination and selectivity of protective devices, to ensure that disconnection times to protect against electric shock and thermal effects are maintained in the event that a supply side device operates; or
(b) the inclusion of monitoring to detect faults that operate supply-side devices, and prevent switchover to island mode operation.

9.6 Electromagnetic compatibility

Electromagnetic compatibility may be affected by a change in earthing system. For example, if an installation contains large quantities of information technology and similar equipment, a TN-C-S or TN-S type earthing arrangement would be anticipated, along with an earthing and bonding system conforming to the recommendations of BS EN 50310. This approach should remain in island mode operation to maintain electromagnetic compatibility.

9.7 Surge and lightning protection

Particularly when operating in island mode, with grid supply disconnected, telecommunication services entering the premises may not be connected to the same earthing system. As a result, the impact of surges due to lightning and HV network switching could well be increased.

Where telecommunication services are present and the system is to be operated in island mode, surge protection is recommended, both on the electrical installation and on any external copper communications cables entering the installation. Relevant standards that should be consulted in providing appropriate surge protection include: BS 7671, the BS EN 62305-series and the BS EN 50174-series.

Where the installation contains renewable generation, such as wind, that may be more susceptible to direct or indirect lightning strikes because of its height in relation to surrounding buildings, lightning and surge protection in accordance with BS EN 62305 and BS 7671 may be required. Further guidance is available in the IET *Code of Practice for Grid Connected Solar Photovoltaic Systems*.

9.8 Inrush currents

Due to the charging of capacitors on the input of many items of equipment, particularly information technology and other electronic devices, large currents can be seen when switching on, or as a result of switching operations in the supply. Where inrush currents are excessive, these may cause the operation of fast-acting protective devices, such as fast-acting fuses and MCBs to the BS EN 60898 series.

The peak inrush current that can be provided is limited by the prospective short circuit current of the supply, or in the case of surges in protective conductor current, the earth fault loop impedance. Inrush currents between phases and between phase and neutral are typically greatest when the grid is supplying power. In TN systems this is also true of protective conductor inrush currents; however, where the grid supply is TT, it may be that protective conductor inrush currents are greatest when operating in island mode.

RCDs may operate due to residual inrush currents, for example, those that charge or discharge capacitors in filters for electromagnetic compatibility, which are often connected between live conductors and the protective conductor within equipment.

BS 7671 requires that consideration is given to nuisance tripping under inrush conditions, and circuits where high inrush currents are typical, such as data centre installations, have measures to help prevent nuisance tripping such as a staggered start of loads.

In certain installations, EESS inverters may not be able to deliver the required currents during switchover, and the designer of the EESS installation should ensure that there are no incompatibilities.

Where problems are envisaged, the options available to designers and manufacturers include:

(a) zero-voltage switching to help limit inrush currents, as the largest inrush currents would otherwise occur when switching at a voltage peak or trough; and

(b) staggering switch-on or switchover of loads during the switching operation. This may require additional switching such as contactors in distribution or final circuits.

9.9 Load handling and load shedding

With smaller capacity EESS, the system may not be capable of powering large loads, and the designer should opt for the system to only supply smaller power loads in island mode. Examples of how this could be achieved include:

(a) disconnecting higher power loads using contactors on individual circuits

(b) having separate distribution boards/consumer units for the lower power 'critical' loads, and only supplying these when in island mode. Arrangements of this type for domestic or small commercial installations are illustrated in Figures 9.8 and Figure 9.9.

9.10 The EESS supplying DC circuits

9.10.1 Powering DC loads

Powering DC loads through the inverter is less efficient, since there are losses in the inverter, and also in the AC to DC converters used to power DC loads from the AC supply.

There is opportunity to minimize these losses by using the battery DC supply to power DC loads. Ideally, loads will be rated for the same voltage as the battery; however, the range of voltage on the battery supply between maximum charge voltage and the minimum healthy discharging voltage can be over 30 %. Where loads are intolerant of this wide variation in voltage, it is necessary to utilize DC to DC converters to ensure a stable supply is available to the DC loads. Despite operating in a similar manner to switched mode AC to DC power supplies, modern DC to DC converters are more efficient and do not exhibit the same losses from rectification. Designers should, however, consider the inefficiencies introduced by resistive losses in ELV DC power circuits where ELV DC is used.

Switchgear and protective devices must be carefully selected to ensure that they are suitable for DC use at the relevant voltages. Switchgear used for LV AC systems will often not be suitable, as they may not be capable of handling DC arcs drawn when contacts are open, and these arcs are more difficult to extinguish than AC arcs. RCDs are not widely available at present for DC systems.

9.10.2 ELV DC supplies

With regards ELV supplies, the issue of voltage drop requires careful consideration in order to avoid excessive losses in wiring and to prevent the voltage at the point of use being too low. For general ELV supplies, the voltage-drop requirements in BS 7671 should be used.

Power for power, ELV supplies require larger cross-sectional area conductors than LV supplies, and this is compounded by voltage drop.

The model of 'fault of negligible impedance' used with LV systems does not always apply with ELV DC systems, often resulting in local heating at the point of fault. Purely electronic supplies may incorporate various forms of electronic overcurrent and/or overvoltage protection. However, this protection should be used in conjunction with suitably selected overcurrent protective devices, such as fuses or circuit-breakers, in case electronic protection fails.

Further guidance in the design and specification of ELV DC supplies can be found in the IET *Code of Practice for Low and Extra Low Voltage Direct Current Power Distribution in Buildings* and the IET Technical Briefing *Practical considerations for d.c. installations*.

9.10.3 LV DC supplies

LV DC supplies are becoming more common for powering information technology and similar equipment, for example, in a data centre or telecommunications environment.

LV DC supplies require earthing in the same way as LV AC supplies. However, care must be taken to prevent electrolytic degradation of the means of earthing and buried metallic services.

Further guidance in the design and specification of LV DC supplies can be found in the IET *Code of Practice for Low and Extra Low Voltage Direct Current Power Distribution in Buildings* and the IET Technical Briefing *Practical considerations for d.c. installations*.

9.11 Isolation and switching off for maintenance

9.11.1 General

Parts of the installation not capable of being isolated by a single device will require a warning notice in accordance with Regulation 537.1.2 of BS 7671. All isolators should have their function clearly marked in accordance with Regulation 537.2.7 of BS 7671.

9.11.2 Dwellings

In a dwelling, BS 7671 requires main isolators to disconnect all live conductors (that is, both line and neutral conductors in single-phase systems, and all line and neutral conductors in three-phase systems). Circuits should be arranged so that it is clear how to remove power from specific final circuits or appliances, as residents and homeowners will be undertaking tasks such as replacing lamps or luminaires, and the isolation of faulty appliances – tasks which might be carried out by skilled or instructed persons in commercial or industrial premises.

9.11.3 Other premises, and work on all electrical installations containing EESS

The Electricity at Work Regulations, 1989 apply and the HSE publication HSG85 *Electricity at work – Safe working practices* provides guidance on these matters. Provisions that the duty-holder should ensure are in place include:

(a) adequate means of isolation that can be identified and locked off;

(b) for each circuit, switchboard/distribution board, and item of fixed equipment, a suitable documented procedure to isolate as required for maintenance purposes; and

(c) competent persons to undertake and manage electrical work, including isolation.

Since an electrical installation containing an EESS will have parts of the installation supplied by at least two sources of energy, isolation procedures for those parts of the installation may not always be simple. It is imperative that those working on the installation can clearly and easily identify the device(s) required to isolate the part(s) of the installation they are working on.

9.12 RCD selection

Where RCDs are used in circuits to be powered by the EESS, either in parallel operation or in island mode, the designer should ensure that they are suitable for the PCE (operating as an inverter) as well as the load.

Where RCDs are used in conjunction with PCE, the RCD shall be selected as illustrated by Table 9.4.

Table 9.4 RCD selection, reproduced from the IET's *Code of Practice for Grid Connected Solar Photovoltaic Systems*

Scenario		RCD selection
1	All PCE contains at least simple separation between AC and DC circuits.	RCD selection need not take into account the PCE.
2	All PCE in the circuit includes at least simple separation or the PCE by construction is not able to feed DC fault currents into the electrical installation.	However, Type B or Type F RCDs may be required depending on the characteristics of loads containing certain power electronic converters such as variable speed drives. These characteristics may differ when the load is fed from PCE rather than the grid.
3	An item of PCE in the circuit does not include simple separation and is able to feed DC fault currents into the installation.	Type B required.

Not all RCDs are suitable for bi-directional applications. Selection of an RCD must take into consideration forward and reverse current flow, where appropriate. RCD manufacturers' information should be consulted.

There are some installation arrangements, for example, connecting vehicle as storage, where two 30 mA RCDs are necessarily connected in series. For instance, where the circuit supplying the EV charging equipment is concealed in a wall at a depth of less than 50 mm, a dedicated Type A 30 mA RCBO is used at the origin of the circuit in the distribution board and the EV charging equipment contains either a Type B RCD or a residual direct current detecting device (RDC-DD), along with a 30 mA Type A or Type F RCD. Selectivity of operating characteristic in this case may not be required, as both RCDs relate to the same dedicated circuit. As two 30 mA RCDs in series will result in either only the upstream or downstream RCD tripping, depending on their tolerances and the actual fault current magnitude, both the upstream and downstream RCD must disconnect all live conductors to comply with Regulation 722.531.3.1 of BS 7671.

Section 10

Network connection and DNO approval

10.1 General

DNOs generally treat the connection of an EESS in parallel with their network in much the same way as they would a distributed generation system, such as a wind turbine or PV system. As such, an installer of an EESS that is intended to operate in parallel with the grid must follow the same approach for gaining connection approval as they would for other types of generation. This requires a good working knowledge of the application of the ENA Engineering Recommendations – in particular G98 and G99 – as described in the following section.

Note: The latest distributed generation guides are available from the ENA website at
https://www.energynetworks.org/electricity/engineering/distributed-generation/dg-connection-guides.html

In applying G98 or G99 to the installation of an EESS on a site with existing generation (or a site where the EESS is being installed together with other generation), the total generation capacity on the site needs to be calculated:

(a) No increase to on-site generation

Where the EESS is DC coupled to an existing generator (for example, linked on the DC side of a PV inverter), there will be no increased inverter capacity on site.

Generally, where an EESS is DC coupled to an existing system, providing that the existing generator has already been notified to the DNO, there is no requirement for the installer of the ESS to notify the DNO.

Note: While, at the time of publication of this Code of Practice, the above statement is true for most DNOs, certain DNOs take a different approach. At the planning stage, the installer should make themselves aware of the notification and approval requirements of the DNO responsible for the installation location. In some circumstances, a DNO may require post installation notification of the system (for systems within the scope of G98), or a pre-installation application (for systems within the scope of G99).

(b) Increased on-site generation

Any site where an AC-coupled EESS is installed, adds 'generation' capacity to that site. For sites without any renewable or other generation, the total on-site generation capacity is simply the rating of the EESS; for other sites it is the combined rating of the EESS and the other generation sources on the site.

For an AC-coupled EESS, the total site generation capacity needs to be determined. This should be calculated as the sum of the nominal AC capacity of the EESS system and the nominal AC capacity of the other generation system(s).

Note: Where PV systems are fitted on domestic properties, it is common for the PV system size to be at or just below the (16 A per phase) G98 limit. Where an AC coupled EESS is installed on these properties, the additional inverter of the EESS will typically take the total installed capacity on the site above 16 A/phase – and hence into G99 territory. Where the total generation is less than 32 A per phase, the G99 fast-track route may be appropriate if all generators are type-tested and the proposed system meets the fast-track requirements.

10.2 Connection requirements for generators

Commission Regulation (EU) 2016/631 Requirements for Generators (RfG) was published in 2016. ENA Engineering Recommendations G98 and G99 have been updated to align with this standard.

Table 10.1 Generator types and DNO EESS notification/approval route

Type	AC grid connection voltage U_0	Generating capacity	Harmonized standard requirements for parallel connection	Total generation on site (A / phase)	EESS capacity (A / phase)	Likely route (Note 1)
A	≤ 230 V	800 W to 1 MW	BS EN 50549-1	≤ 16	≤ 16	G98 (single or multiple premises)
				≤ 16	≤ 32	G99 route 1 or G99 route 2
				> 16	n/a	G99 route 3
					> 32	
	> 230 V to 1,000 V					
	> 1,000 V to 110 kV (Note 3)		BS EN 50549-2 (Note 3)			
B	≤ 1,000 V	> 1 MW to 50 MW	BS EN 50549-1	n/a	n/a	
	> 1,000 V to 110 kV (Note 3)		BS EN 50549-2 (Note 3)			
C	≤ 110 kV (Note 3)	> 50 MW to 75 MW	Note 2			
D	≤ 110 kV (Note 3)	> 75 MW	Note 2			
	> 110 kV (Note 3)	Any	Note 2			

Note 1: For G98, G99 route 1 and G99 route 2, type- tested generation is required.
Note 2: Harmonized standards for generators exceeding 50 MW generating capacity are currently under development.
Note 3: Generators operating at HV are outside the scope of this Code of Practice.

Type A generators are considered the simplest, and hence have the least onerous requirements. As we move through Types B to D, additional requirements are introduced, relating the performance of the generator, and its reaction, to progressively more network-related conditions.

10.3 Engineering Recommendations G98 and G99

Published by the Energy Networks Association (ENA), Engineering Recommendations G99 and G98 set out the connection and commissioning requirements for connecting an electricity generating system to UK distribution networks. Both documents are referenced within the Great Britain Distribution Code. The following flowchart describes the process for identifying the appropriate procedure:

10.3.1 Dwellings

It is recommended that installations in dwellings include type-tested generation (inverters). If that is the case, EESS installations in many dwellings will follow one of the following DNO notification/approval routes:

(a) G98 single site or multiple site, where the total generation per dwelling including EESS is 16 A or less per phase; or

(b) G99 route 1 or route 2, where the total generation per dwelling is 32 A or less per phase, and the EESS generation is 16 A or less per phase.

Section 10 – Network connection and DNO approval

10.3.2 Larger installations

Larger installations, perhaps typical of commercial or industrial applications, will follow one of the G99 routes.

Figure 10.1   G98 versus G99 flowchart

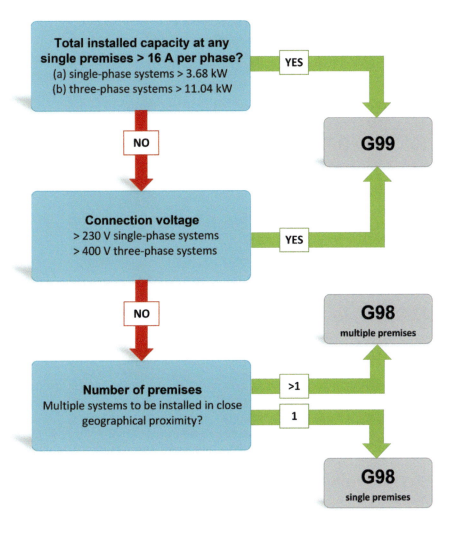

Section 10 – Network connection and DNO approval

10.4 G98: single premises

The process for installing a single system under Engineering Recommendation G98 is a very straightforward 'fit and inform' procedure, with DNO notification only required after the installation has occurred.

Figure 10.2   G98 single site process

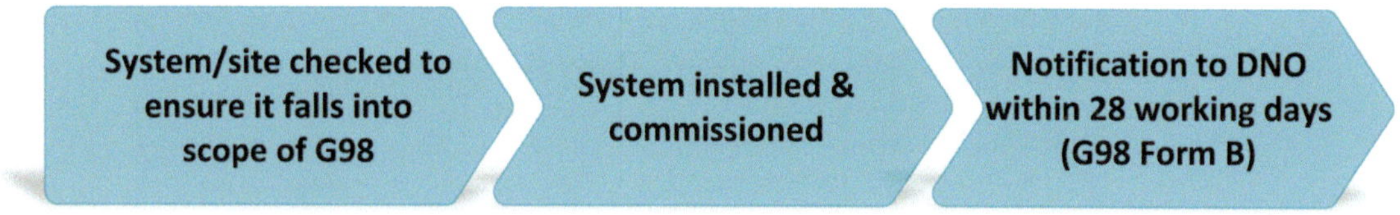

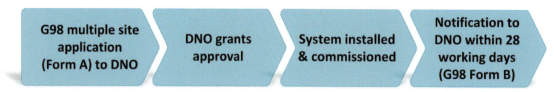

To comply with the simple G98 fit and inform process, the site shall:

(a) have a total site aggregate generator/inverter capacity of less than or equal to 16 A/phase.

(b) be a single installation on a single property.

(c) use G98-type approved inverters. A list of approved inverters is available from the Type Test Register published on the ENA website at https://www.ena-eng.org/gen-ttr/.

(d) be installed in line with the connection and commissioning requirements of Engineering Recommendation G98.

(e) be notified to the DNO within 28 days using the Engineering Recommendation G98 commissioning form.

Note: Notifying the DNO is a legal requirement under the ESQCR 2002.

10.5 G98: multiple premises

Due to the potentially larger impact on the electricity network, when installing multiple PV systems within a close geographic region, approval from the DNO is required before installation proceeds. Within Engineering Recommendation G98, the definition of close geographical proximity is taken to be where the postcode, excluding the last two letters, is the same (for example, CA11 0xx); or where any two planned installations are within 500 m of each other. Figure 10.3 illustrates the process.

Figure 10.3   G98 multiple site process

As with the single site G98 process, each system shall:

(a) have a total aggregate AC capacity of less than or equal to 16 A/phase;

(b) use Engineering Recommendation G98 approved inverters;

(c) be installed in line with the connection and commissioning requirements of Engineering Recommendation G98; and

(d) be notified to the DNO within 28 days, using the Engineering Recommendation G98 commissioning form.

Note 1: For systems less than 16 A/phase there are no requirements to balance systems across phases. However, for multiple installation projects, such as those on housing estates, agreement on phase balancing will need to be reached with the DNO.

Note 2: Notifying the DNO is a legal requirement under the ESQCR 2002.

Note 3: A list of G98 type approved inverters can be found here: https://www.ena-eng.org/gen-ttr/

10.6 G99: installations

Systems of over 16 A per phase need to follow the process that is set out in Engineering Recommendation G99. The procedure is slightly different depending on the size and type of the system (see Figure 10.4).

Figure 10.4 G99 process

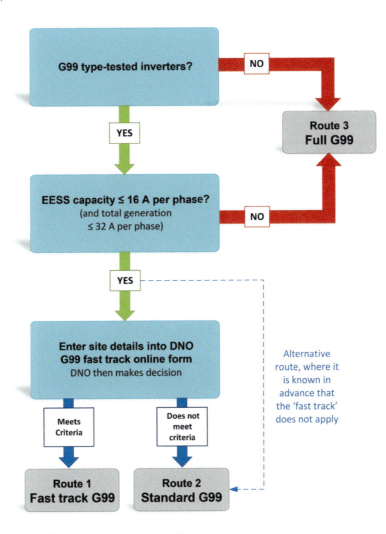

10.6.1 G99 Fast track process (route 1)

The fast track process has been created to provide a simplified connection process for a small-scale EESS that is being installed in addition to existing site generation. The major application for this process is seen as an EESS added to an existing domestic PV system.

The fast track process gives an instant response, incurs no fees and requires no site witness testing. The process is as follows:

(a) customer applies for connection using the DNO's online application form;

(b) if all the criteria are met, the customer is given instant approval to connect;

(c) commissioning must be at least 10 working days from the date of application, but not more than 3 months in advance; and

(d) within 28 days of commissioning, applicant uses a web-link to upload commissioning sheets for the EESS scheme.

Section 10 – Network connection and DNO approval

Due to small variations between DNO networks, it is necessary for each DNO to have its own fast-track web application form. Like any part of the G98 or G99 process, it is obviously key that a customer applies to the appropriate DNO (however, the application will be rejected if the site MPAN and DNO do not match). The DNO can be identified either geographically, or from the distributor ID from the MPAN (see Figure 10.5). Lists of DNOs' distributor IDs are available on various industry web-sites.

Figure 10.5 Example of MPAN information

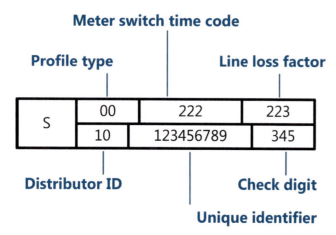

It is to be noted that if the fast-track criteria are not met, the application process reverts to the normal G99 process. Where it is known in advance that the fast-track process will not apply, the standard G99 application process can be followed from the outset.

10.6.2 Normal G99 process (routes 2 and 3)

The normal G99 process requires a written connection application and formal DNO approval prior to proceeding with the scheme. The process varies slightly depending on the size of the EESS and whether the EESS utilizes G99 type-tested inverters.

The use of type-tested inverters simplifies installation and commissioning and generally means that DNO witnessed commissioning tests are not required. A full list of type-tested inverters is available on the ENA website: https://www.ena-eng.org/gen-ttr/

Section 10 – Network connection and DNO approval

The process is described in Table 10.2.

Table 10.2 Normal G99 process

Route 2	Route 3		
EESS with Type A generators ≤ total ≤ 16 A per phase *and* total generation at the site ≤ 32 A per phase *and* **only type-tested inverters are used**	All other systems (Type A generators rated > 16 A per phase, and all Type B, C and D generators)		
G99 application form and EESS form submitted to DNO			
Discussions with DNO where required	Discussions with DNO		
	Final design and planning		
	Contestable and non-contestable work split agreed with DNO (as required)		
Connection offer from DNO			
Offer accepted by client			
Installation and commissioning	Installation		
	Commissioning witnessed by DNO if requested		
G99 commissioning form supplied to DNO within 28 days			

10.6.3 G99: additional form for battery storage systems

A standard form has been developed by the ENA, as an adjunct to G99, to collect information on the installation of an EESS. This form requests the following information about the EESS:

(a) details of storage technology;
(b) whether the request is for storage only or storage combined with another technology;
(c) nameplate power rating of storage (MW);
(d) registered energy storage capacity (MWh);
(e) security of supply required – import/export;
(f) details of operating mode or commercial service being provided; and
(g) details of maximum power ramp rates (import, export and swing) in MW/s.

10.6.4 G99: commissioning

For larger systems, the DNO will typically require the installation of a standalone G99 protection system. The G99 protection is normally commissioned by a series of tests that simulate grid out-of-bounds conditions (under- and overvoltage, etc.). For the duration of the tests, the sense circuits on the G99 protection are temporarily disconnected from the grid and connected to a test set with a variable voltage and frequency output. Commissioning then involves cycling the relay through each of the G99 voltage and frequency settings and timing the response to ensure that the system meets the prescribed G99 values and reaction times.

10.6.5 G99: arrangements for HV connections

For HV connections, the configuration of the G99 protection system will be agreed with the DNO during the design phase. This process will include consideration of the number/location of G99 protection sets (for example, G99 protection may be required for each inverter station); it will also consider what switchgear the G99 protection controls (for example, the relay may monitor the LV side but control HV switchgear).

Note: Whilst HV EESS are outside the scope of this Code of Practice, the DNO supply may be provided at HV.

10.7 Labelling requirements

System labels and signs shall be installed in accordance with the requirements of BS 7671 and Engineering Recommendations G98 and G99. Labelling shall include:

(a) identification of the mains AC isolator (point of emergency isolation).

(b) system schematic displayed at point of interconnection with the DNO's distribution system. The diagram shall include information on the owner and operator of the system and a summary of the protection settings.

(c) dual supply label at the service termination, meter position and all points of isolation between the EESS and the main incoming supply.

10.8 New connections

The process for getting a new connection made to the electricity network is broadly similar to that of getting connection approval in accordance with Engineering Recommendation G99. The steps for gaining a new connection to a DNO's network are as follows:

(a) notify the DNO as early as possible, preferably at the initial planning and design phase, of the details of the scheme to be connected, the time the connection is required and the maximum capacity (import and export) needed from the connection.

(b) the DNO will provide a quotation for the costs of connecting to its distribution system.

Note: When providing a quotation it is common practice for the DNO to advise the customer that it needs to nominate a supplier before the connection can be made and preferably before accepting the quotation.

(c) connection offer accepted and connection agreement signed.

(d) connection made – the DNO maintains the connection.

Alternatively, a developer may choose to approach an independent connection provider (ICP) to arrange and provide the connection. The ICP must be registered and approved by the DNO to whose network the connection will be made. While the ICP may carry out the connection to the DNO network, they do not have distribution licences, so will not be licensed to operate the connection once it is installed (the DNO will adopt the asset on completion).

Note: Further information can be found in the ENA Competition in Connection Code of Practice: http://www.connectionscode.org.uk/assets/files/CiC_CoP_v1.5_October_2019.pdf

10.9 Contestable and non-contestable works

The works to connect to the grid are often made up of two types:

(a) contestable: work, typically on new infrastructure, that either the DNO or an independent company can provide; and

(b) non-contestable: work performed by the DNO on existing network equipment owned by the DNO.

Contestable work is done to the DNO's specifications but may be tendered to allow for the most financially competitive solution. Contestable work typically includes:

(a) the design for the network extension and any contestable reinforcement of the existing network;

(b) procurement of equipment and materials to the DNO's current specification for the extension;

(c) trenching and other preparation of the site;

(d) constructing the network extension and any contestable reinforcement and diversions;

(e) drawings and site records;

(f) reinstatement works; and

(g) making provision for the installation of metering equipment.

Section 10 – Network connection and DNO approval

Typical non-contestable elements of work that can only be provided by the DNO include:

(a) determining the point of connection to the distribution system;

(b) approving contestable designs for new connections/reinforcements;

(c) planning, designing, specifying and carrying out any non-contestable works;

(d) removing or repositioning existing electrical plants and electric lines;

(e) connection to the distribution system; and

(f) operating, repairing and maintaining the electrical plant.

10.10 Engineering recommendation G100

An export limitation scheme measures the active power at key points within the customer's installation and then uses this information to either restrict generation output and/or balance the customer's demand (for example, by charging an EESS) in order to prevent the export to the distribution system from exceeding the agreed export capacity. In some circumstances, an export limitation system may include a secondary feature to restrict generation export when the voltage at the connection point exceeds the statutory voltage limits.

Engineering Recommendation G100 *Technical Requirements for Customer Export Limiting Schemes* defines the technical design requirements for export limitation schemes that limit the net site export to below an agreed maximum and are installed on the customer's side of the DNO's connection point. The technical requirements within G100 include a maximum response time of 5 seconds, interconnection specifications (where the system is constructed from discreet components), accuracy thresholds and a requirement for fail-safe operation.

As well as describing the technical requirements of an export limiting scheme, G100 also details the tests expected by a DNO during the commissioning of an export limitation system to prove the correction function of the system. A commissioning form is provided within G100.

G100 should always be read in conjunction with G98 and G99.

Export limiting arrangements are a common feature of many EESS schemes - either as a means to divert power when export is being limited or because the EESS is larger than the site export capacity. Hence G100 is an important document for many EESS schemes.

EESS Installation

11.1   Installation phases

Figure 11.1   Typical phases of installation and commissioning

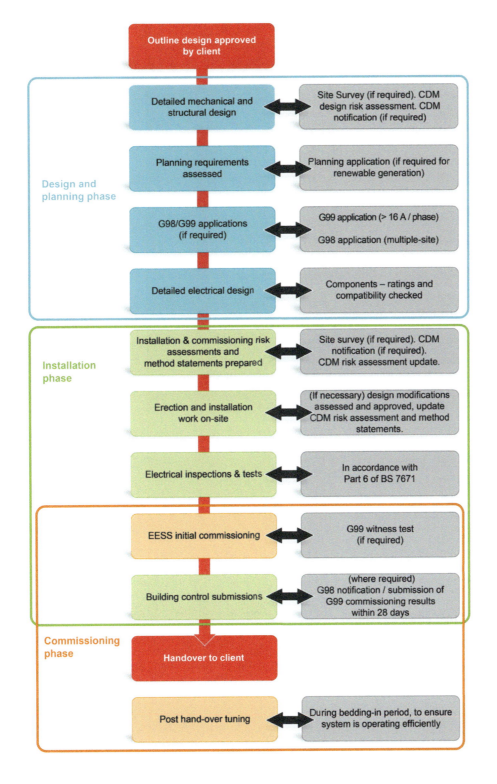

Section 11 – EESS Installation

11.1.1 Design and planning phase

The design and planning phase is essential before final procurement activities occur and physical work begins on site.

In addition to assessing the purpose and operational modes/needs of the EESS, various permissions and notifications for the building/construction work are addressed:

(a) all work on new electrical installations, as well as additions to existing installations, is classified as construction work, and relevant parts of the CDM Regulations apply. Larger projects will require notification.

(b) certain aspects of the CDM designer's duties apply for all projects, notifiable or otherwise. For domestic customers, some of the duties of the client fall to the principal contractor. It is important for designers to be familiar with EESS and the type of installation it is to serve.

(c) planning provisions must be met where EESS is located outside the building and formal planning permissions may be required for larger EESS.

(d) even where planning permission is not required, the relevant provisions for safety and functionality of the technical requirements of the Building Regulations apply. In respect of EESS, this may include provisions for fire, ventilation, energy efficiency and electrical safety. Under the appropriate Building Regulations certain works on electrical installations require to be notified to the relevant local authority building control department. These notification and documentation requirements must be met. The provision of new (rather than replacement) EESS is likely to fall into this category. Depending on how an EESS is replaced, system replacement work may also be classed as notifiable work.

(e) where appropriate, Engineering Recommendations G99, G98 or G100 applications should be submitted.

(f) finally, the electrical design should be completed.

11.1.2 Installation phase

The installation phase begins when final preparations are being made to commence work on site. Relevant risk assessments will be in place to cover health, safety and welfare during installation and commissioning.

Installation activities will be carried out to a high standard using industry-approved practices. Trained and competent individuals should be employed.

Any design changes will be approved and documented in accordance with the relevant processes to ensure that:

(a) the design as installed continues to meet relevant requirements and legislation; and

(b) any as-built documentation accurately reflects what is installed on site.

Accuracy of as-built operation and maintenance information is particularly important with EESS and other embedded generation schemes, as safety issues such as location and function of isolators is paramount for the safety of those maintaining and/or decommissioning the electrical installation after handover.

11.1.3 Commissioning phase

Commissioning is discussed in more detail in Section 12. The commissioning phase commences before installation is complete so that system setup and test operation of the relevant controls, switching, etc. can begin.

Commissioning for some systems will not be completed until the system has been in use for a period of time and final 'tuning' is addressed. Some commissioning, for example, where the system includes renewable generation, may only be possible at certain times of the year or under certain weather conditions.

11.2 Battery installation

In determining an appropriate location and means of installation for a battery, a number of hazards need to be considered.

11.2.1 Hazards

11.2.1.1 Battery component substance

Batteries by their nature often contain hazardous substances and the following legislation may apply in their transportation, use and disposal:

(a) Waste Electrical and Electronic Equipment Regulations 2013; and
(b) Batteries and Accumulators Regulations 2009.

The hazards vary depending on the type of battery and the results of inappropriate chemical contact or environmental contamination may be severe. Manufacturer's safety data sheets (MSDS) need to be consulted over the chemical hazards contained within a battery. The MSDS will describe the hazards the chemicals contained in the battery may present, and will give information on the handling, transportation, storage and emergency measures in case of accident. See also Section 11.2.1.5.

As an example, components of lead-acid type batteries may be more hazardous than other technologies such as lithium-ion. Lead is a toxic metal that is harmful to human health and the environment, and the battery type also contains strong acids that are highly corrosive and may permanently damage skin, eyes and other organs.

Corrosion-resistant bunding or enclosures may be necessary for some battery types and are recommended so that ground water and other environmental contamination for all batteries containing liquid or gel chemicals can be avoided.

11.2.1.2 Battery gassing

BS 7671 requires that adequate ventilation is provided for stationary batteries.

Charging some batteries, particularly wet-cell lead-acid batteries, can result in the production of gases such as hydrogen and oxygen that can in turn result in an ignition and explosion risk. Charging of a lead-acid battery produces hydrogen, which, being lighter than air, will gather in the upper part of a battery enclosure/room.

The gases emitted during charging may also contain a mist of the chemicals within, such as sulphuric acid, which can be a hazard to breathing and to eyes.

Suitable fresh air ventilation is required for batteries that may evolve hydrogen on charging. In accordance with HSE Publication INDG 139, the ventilation should ensure the hydrogen concentration is kept well below:

(a) 1.0 % in all enclosures; and
(b) 0.4 % in rooms/locations where people may enter.

Section 11 – EESS Installation

HSE Publication INDG 139 and BS EN IEC 62485-2 outline the concentration limits for safety. A hazardous area assessment is required to determine the hazardous area zone around the battery in accordance with the Dangerous Substances and Explosive Atmospheres Regulations 2002 and BS EN 60079.

11.2.1.3 Electrical short circuit

A large amount of energy can be stored within a battery, making the accidental short circuiting of the battery terminals a potential hazard. This can result in large fault currents with the risks of burns and explosion.

Protection of the battery terminals by barriers or insulation is a key consideration in preventing accidental short circuits, and is a recommended precaution even where the battery is installed in enclosures. The terminals need to be protected during normal operation and care needs to be taken to prevent shorts during installation and maintenance (for example, by the use of insulated tools and the removal of metallic jewellery, necklaces etc. before carrying out work).

BS 7671 requires terminals of batteries to be arranged such that inadvertent simultaneous contact with voltages in excess of 120 V is prevented.

Even where all practicable precautions are considered during the design of the installation or equipment, batteries cannot be made fully 'dead' before removal. It is therefore necessary for all those working with batteries to utilize a safe system of work to reduce and eliminate electrical risks and hazards associated with batteries. Battery installations should be designed with safety in mind.

Note: HSE guidance on working with batteries is available in publication INDG 139.

Safe systems of work for battery installations should consider:

(a) removal of jewellery, metal-framed glasses on cords, and articles of conductive clothing;

(b) use of insulated tools, and appropriately fused test equipment with test leads complying with BS EN 61010-031/HSE Guidance Note GS 38;

(c) use of insulating terminal covers and shrouds when leads and batteries are disconnected;

(d) procedure for removal of barriers covering terminals; and

(e) provision of appropriate personal protective equipment (PPE), particularly on larger systems with high discharge energies, such as protective clothing, gloves and face shields to guard against arcing and fire.

All those using safe systems of work and PPE should be appropriately trained.

11.2.1.4 Fire

Lithium-ion batteries can be subject to thermal runaway, which, in extreme cases, can result in cell rupture, explosion and fire. Lithium-ion battery packs contain control circuitry, which protect the battery against these hazards.

Other battery chemistries have the potential for rupture and fire.

Care needs to be taken to ensure that:

(a) manufacturer's installation guidance is followed;

(b) control circuits are not damaged or modified;

(c) battery terminals are not shorted or connected in reverse polarity;

(d) batteries are only connected to approved charge/discharge units;

(e) battery cases are not damaged or punctured; and

(f) batteries are not subject to excessive heat.

Section 11 – EESS Installation

Batteries should have a suitable control circuit and be matched with an appropriate charger to avoid explosion risks.

11.2.1.5 Weight, manual handling and transportation

Batteries can be heavy, and consideration of manual handling operations for installation, maintenance and decommissioning is important.

The weight of the batteries, and any associated enclosures, racks or stands, also needs to be considered with respect to mounting: floors, walls and fixings must be suitable for the weight they are to support.

Larger batteries can have a substantial weight and will need a robust base – typically a ground floor concrete pad. Mounting such batteries on upper stories or in a loft should generally be avoided. For very large batteries, point and distributed loading needs to be addressed with structural engineers. Larger batteries will require larger and more robust enclosures, racks or stands.

The Carriage of Dangerous Goods and Use of Transportable Pressure Equipment Regulations 2009 and the European Agreement concerning the International Carriage of Dangerous Goods by Road (also known as the ADR) both apply to the transportation of batteries. Transport of many kinds of battery is regulated, for example:

(a) lead-acid batteries are regulated as dangerous goods under Class 8 (corrosive substances), and also because of the risks associated with fire and explosion if their terminals are shorted.
(b) lithium batteries are also regulated and their transport must comply with Class 9 Directives, due to fire and explosion risks. They may only be transported after passing UN 38.3 tests.

11.2.2 Installation requirements

Information on the safe installation of batteries is contained in BS EN IEC 62485 *Safety requirements for secondary batteries and battery installations*.

Information relating to system components integrated with solar PV systems is also provided in the *IET Code of Practice for Grid Connected Solar Photovoltaic Systems*.

11.2.3 Battery location

When determining where a battery is to be located, the following factors need to be considered:

(a) manufacturer's instructions and SDS;
(b) limits on the length of cable between battery and charger/inverter, where these are located in separate enclosures;
(c) space for cables and containment to be routed, taking into consideration their minimum bend radii;
(d) whether workspace and access to equipment is adequate, so as to prevent electrical and other dangers during installation, maintenance and decommissioning, including arc flash hazard safe distances (see Appendix E) and any safe working distances recommended by equipment and battery manufacturers;
(e) whether the means of isolation and other protection is clearly identifiable and readily accessible to the persons who need to access it;
(f) ventilation requirements;
(g) ambient temperature range;
(h) distance to other heat sources;
(i) location with respect to fire escape routes and exits;
(j) presence of fire detection systems;
(k) presence of sources of ignition (such as gas boiler pilot light);

Section 11 – EESS Installation

(l) weight – batteries may be very heavy and the arrangements for mounting on the building fabric must therefore take into account the capability of the fabric to support both the distributed and point loading for the weight of the batteries; and

(m) flood risks.

While good guidance is available for the installation of lead-acid batteries (for example, BS EN IEC 62485 *Safety requirements for secondary batteries and battery installations*), there is currently no equivalent for lithium-ion batteries.

The impact of the battery on fire safety, fire risk assessments and compliance with Part B of the Building Regulations may need to be considered, and, for some battery types, advice should be sought from specialists.

It is not practical to provide hard and fast rules on battery locations within buildings, however, Table 11.1 highlights some of the considerations for some of the key locations currently used to locate lithium-ion batteries in a dwelling.

BS EN 547-series can be useful in helping to determine the suitability of access provisions for the installation, commissioning, maintenance and decommissioning. Certain factors, such as the need to maintain a means of access/egress and escape for safety, will take precedence over these standards, and may require greater dimensions.

BS 7671 requires that conductive parts of stationary batteries are located so that they are accessible only to skilled or instructed persons, i.e. the battery is installed in a secure location or, for small batteries, a secure enclosure is provided.

Table 11.1 Examples of considerations for battery locations in a dwelling

Location	Considerations
Loft	• Can the battery tolerate the high temperatures present in the loft over the year?
	Note: While loft temperatures may not present a risk of triggering thermal runaway, they may be out of normal operating range and could reduce battery life span.
	• Can the battery tolerate the low temperatures present in the loft over the year (many batteries perform less well at low or sub-zero temperatures)?
	• Is suitable fire detection present in the loft?
	• Is suitable unimpeded access available for installation and maintenance?
	• Are there access provisions for lifting heavy batteries into and out of the loft area and the enclosure?
	• Is there suitable access to emergency shutdown apparatus?
	• Are loft joists suitable for additional weight?
	• Is there adequate lighting to carry out all the required maintenance activities?
	• Is there adequate ventilation where required for the particular battery chemistry?
Under kitchen or utility-room cabinets	• Can the battery tolerate the increased temperature from adjacent appliances (such as ovens and fridges)?
	• Is the battery access suitably restricted to authorized personnel?
	• Is the battery suitably ventilated?
	Note: Kitchens often contain ready sources of ignition and may not be a suitable location if highly explosive gases, such as hydrogen, may be generated by the battery.
	• Is the battery adequately protected against the effect of liquid spills, and any heat generated?
	• Is there adequate ventilation where required for the particular battery chemistry?

Table 11.1 *Cont.*

Location	Considerations
Under stairs cupboard	• Does the location compromise fire escape routes? • Is the battery access suitably restricted to authorized personnel? • Is the battery suitably ventilated? • Is suitable fire detection present under the stairs? • Is there adequate ventilation where required for the particular battery chemistry?
Garage	• Can the battery tolerate the ambient temperature range (especially in detached outbuildings or garages, the high temperatures in summer and low/sub-zero temperatures in winter may lead to reduced battery performance)? • Is suitable fire detection present? • Is there adequate ventilation where required for the particular battery chemistry?
Outside	• Can the battery tolerate the ambient temperature range (high temperatures in summer and low/sub-zero temperatures in winter may lead to reduced battery performance)? • Are battery components suitable for outdoor installation (appropriate IP ratings)?

11.2.4 Fire detection and alarm

Where a system has inverters or switchgear installed in a loft (or other similar rarely visited building zones), it is recommended that appropriate fire detection equipment should be installed in the loft (or other rarely visited zone). Where sensors are mounted in a building location where an alarm may not be heard (for example, in a rarely visited zone) it will be necessary to link the sensor to another sounder or a sensor that will sound elsewhere in the building.

Where an EESS system is installed in commercial premises, the EESS installer shall bring to the attention of their customer the need to arrange for a review of the fire risk assessment.

Where an EESS system is installed in domestic premises, the EESS installer shall bring to the attention of their customer the strong recommendation for having suitable fire detection in any loft (or other rarely visited zone) in which batteries or PCE are installed. It may be necessary for such work to be completed before EESS is installed, in particular, where a Category LD1 or LD2 system is installed. Domestic smoke and fire detection systems should be installed to the requirements of BS 5839-6:2019. The standard addresses this specific requirement:

(a) Clause 11.1.1 requires Category LD2 systems to incorporate a smoke or multi-sensor detector in lofts. (This was a recommendation in the previous edition of the standard.)
(b) Clause 11.2 g) recommends smoke or heat detection to be installed in lofts in Category LD1 systems.
(c) for any category of fire detection and alarm system, Clause 11.2 p) recommends smoke, heat, or multi-sensor detection/alarm where PV power systems, boilers and UPS systems (which would include EESS) are installed in loft spaces.

Note: In BS 5839-6:
  (a) a Category LD1 system is a system installed throughout the dwelling, incorporating detectors in all circulation spaces that form part of the escape routes from the dwelling, and in all rooms and areas in which fire might start, other than toilets, bathrooms and shower rooms; and
  (b) a Category LD2 system is a system incorporating detectors in all circulation spaces that form part of the escape routes from the dwelling, and in all rooms or areas that present a high fire risk to occupants.

In systems with combined detector and sounder alarms (Grades D1, D2, F1 and F2), sensors or alarms installed in spaces that are infrequently visited, such as loft spaces, should be interlinked with those in other parts of the premises.

11.3 PCE/BMS location

PCE and BMS need to be mounted in a suitable location, taking into account factors such as:

(a) accessibility for installation, maintenance and decommissioning;

(b) accessibility for any controls and isolation;

(c) cable routes;

(d) environmental aspects against the ingress protection (IP) rating – where the inverter is to be located outdoors, it must have a suitable rating such as IP 44 or IP 54;

(e) heat dissipation and ventilation;

(f) the impact of any noise generated by the PCE (some PCEs may have audible noise from fans and electronic components and, on rare occasions, noise outside the audible range may cause a problem for pets and other animals);

(g) proximity to batteries (there may be limitations in voltage drop or cable size that limit the distance between the batteries and inverter);

(h) weight (the arrangements for mounting on the building fabric must take into account the capability of the fabric to support both the distributed and point loading for the weight of the inverter);

(i) orientation (the arrangements of ventilation and cooling in the inverter may dictate not only the proximity to walls and other devices, but also the orientation to enable convection to assist or effect cooling); and

(j) proximity to audio frequency induction loop system (AFILS) components (AFILSs may be impacted by excessive audio-frequency, ultra-sonic and electromagnetic noise).

11.4 Combined EESS units

In some cases one enclosure may contain both the batteries and the charger/inverters. Such units will be installed taking into consideration the issues outlined in both Sections 11.2 and 11.3.

Section 12

EESS inspection and testing

The commissioning process for an EESS shall include:

(a) inspection and testing of the system to the requirements of BS 7671;

(b) the commissioning procedures required by the DNO under Engineering Recommendations G98/G99 – as relevant; and

(c) the commissioning requirements as specified by the manufacturer(s) of the EESS components.

Inspection should be performed prior to testing and normally prior to energizing the installation. Testing should generally be performed in the prescribed sequence. In the event of any test indicating a fault, once the fault has been corrected, preceding tests that may have been influenced by the fault should be repeated.

The inspection and testing of the AC circuit(s) of an EESS can be performed in exactly the same manner as that applicable to any other AC circuit. The procedure is described in BS 7671 and supporting documents, in particular Guidance Note 3 *Inspection and Testing*.

The commissioning procedures required by the DNO are set out in Engineering Recommendations G98/G99.

In addition, where an EESS is DC coupled to a solar PV system, the inspection and testing of the DC side of the system needs to follow the requirements of BS EN 62446-1.

12.1 Inspection requirements

The inspection of an EESS installation needs to follow the requirements of BS 7671 and the manufacturer's instructions for the components forming the EESS.

Table 12.1 describes the general items to be inspected and should be applied to the whole installation (both AC and DC sides).

Note: BS EN 62446-1 provides an additional inspection checklist for solar PV systems.

Table 12.1 Inspection checklists

A	System design/equipment selection and connection
	General check to confirm equipment complies with relevant standards.
	System installed in accordance with design.
	Equipment correctly selected and erected – according to standards and manufacturer's instructions.
	Equipment undamaged.
	Conductors correctly connected, identified and routed.
	Conductors suitably selected for voltage and current requirements.
	Switchgear and ancillary devices suitable for voltage and current requirements.
	Switchgear and ancillary devices correctly installed and connected.
	Components selected and erected to suit the location and any external influences.

Table 12.1   *Cont.*

B	Protection against electric shock
i	*Basic protection (prevention of contact with live parts)*
	Protection by insulation of live parts (e.g. conductors) – correctly applied and suitable.
	Protection by the use of barriers or enclosures – suitable and with an appropriate IP rating.
ii	*Additional protection (additional protection measures)*
	Protection by the use of an RCD not exceeding 30 mA.
	Supplementary bonding.
iii	*Other methods of protection*
	SELV.
	PELV.
	Double or reinforced insulation (Class II or equivalent equipment and circuits).
iv	*Automatic disconnection of supply*
	Suitable main earth provided.
	Earth connections suitably installed.
	Main protective bonding suitably installed.
	RCDs provided for fault protection.
C	**General**
	Equipment and circuits deployed to prevent mutual detrimental influence.
	Equipment located to allow suitable access arrangements.
	Labels and signs displayed and durable.
D	**Systems that provide back-up or island mode power supply**
	Isolation of grid supply current-carrying conductors (see Section 9.4).
	Provision of consumer's earth electrode (see Section 9.4.10).
	Additional protection by RCDs to supplement overcurrent protective devices to ensure adequate disconnection times are met (see Section 9.4.11).
	Provision of surge protection devices (see Section 9.7).
	Staggered start and/or load-shedding on change-over (see Sections 9.8 and 9.9).
E	**EESS location**
	Location/room has ventilation arrangements suitable for EESS and ancillary equipment.
	Location selected suitable for IP rating of EESS components.
	Fire prevention and detection measures implemented – as necessary to meet EESS technology.
	Site safety signs displayed and durable – as necessary to meet EESS technology.
F	**EESS equipment**
	Equipment installed to manufacturer's requirements.
	Clearance from adjacent objects meets manufacturer's specifications.
	Ventilation of equipment meets manufacturer's requirements and remains unobstructed.
	Clearance from any external heat sources acceptable – as relevant.
	EESS equipment suitably accessible for maintenance and operation.
	EESS labelling displayed and durable (e.g. 'dual supply' signs).

Section 12 – EESS inspection and testing

12.2 Circuit test requirements

12.2.1 General

The testing of an EESS installation needs to follow the requirements of BS 7671, the requirements of ER G98/G99 as relevant and the manufacturer's instructions for the components forming the EESS.

Unless specified by the manufacturer(s), circuit testing generally only needs to be done on site-installed circuits, i.e. those circuits formed by the installer during construction works. Pre-wired circuits forming part of the equipment supplied by the manufacturer are generally out of the scope of on-site testing, unless the manufacturer specifies otherwise.

Note 1: Circuit test requirements apply to all parts of the system – i.e. both AC and DC sides.
Note 2: BS EN 62446-1 describes additional tests for solar PV systems.

Should any test indicate a fault, once the fault has been corrected, that test, and all preceding tests that may have been influenced by the fault, should be repeated.

Table 12.2 summarizes the minimum test requirements for the electrical circuits forming an EESS. Instructions for performing the tests are provided in BS 7671 and supporting documents, in particular Guidance Note 3 *Inspection and Testing*.

12.2.2 Determination of earth fault loop impedance and prospective fault current

It should be noted that earth fault loop impedance and prospective fault current measurements may be distorted by the presence of local inverters (from EESS, solar PV, and so on). This may be due to the following:

(a) when testing in island mode an EESS or solar PV inverter may appear to act as a constant current source, and indicate a far higher than usual prospective fault current and a far lower than usual loop impedance value, if tests are carried out at currents below the sum of the current limits of the inverters connected at the time of the test. If these tests are carried out in island mode, particularly with a small EESS, the tests may not provide an adequate reading before inverter current limit protection operates.

(b) when testing grid-connected inverters operating in connected mode, the inverter output tracks the change in supply voltage, so the contribution to prospective fault current from the inverter may not be detected by the test instrument.

It is therefore strongly recommended that:

(a) external earth fault loop impedance and prospective fault current measurements are carried out at the origin, with inverters isolated.

(b) earth fault loop impedance and prospective fault current measurements for the remainder of the installation are not taken unless all the inverters in the system are isolated. In order to provide verification for automatic disconnection of supply, the following tests can be used:

    (i) for connected mode: the verification of earth fault loop impedance (Z_{DB} and Z_S) is based on the external earth fault loop impedance, Z_e, plus measured (R_1+R_2) values. Prospective fault current should be the measured prospective fault current plus the sum of the current limit of all connected-mode inverters.

    (ii) for island mode: if applicable, verification of earth fault loop impedance (Z_{DB} and Z_S) is determined using measured (R_1+R_2) values, plus the manufacturer's information regarding the value of Z_e to be assumed for the EESS or the relevant PCE within it. Prospective fault current should be the sum of the current limit of all connected-mode inverters.

Section 12 – EESS inspection and testing

Table 12.2   Minimum test requirements for electrical circuits

Item	Requirement
Continuity of earthing and/or equipotential bonding conductors	• Test to ensure the continuity of earthing and/or equipotential bonding conductors (if fitted). • Connection to main earthing terminal verified.
Insulation resistance test	• Test to determine the insulation resistance between live conductors, and between live conductors to protective conductor connected to the means of earthing. • Test to be undertaken on all site-installed circuits (AC and DC). **Note:** Devices that may influence, or be damaged, during insulation resistance testing (e.g. SPDs) should be temporarily disconnected during testing.
Polarity test	• Test to ensure that all circuits are correctly marked and connected.
Earth electrode resistance	• Test to determine the electrode resistance(s) to earth (where fitted). **Note:** Some installations may have more than one earthing system, each with earth electrodes (see examples in Figure 9.11, Figure G.6 and Figure G.7). In these cases, the earth electrode systems should be verified separately.
Earth fault loop impedance	• Determine the earth fault loop impedance of individual circuits. • The determined value needs to be checked to ensure it is low enough to allow the circuit protective device to clear a fault within the proscribed disconnection time for that circuit. **Note:** See Section 12.2.2
RCD test	• Test to ensure proper function of any RCDs. **Note:** Where the RCD is tested from the downstream line(s) to the associated protective conductor, the test should be carried out in connected mode and island mode to ensure that the earth fault return path is operational in both modes.
Functional tests	• Tests to ensure switchgear and other circuit devices operate correctly. • Tests to ensure correct functioning of island mode isolator in switching between all supported modes.

12.3 EESS Commissioning

The procedure for commissioning an EESS will depend on the type of system, the use of the system and the requirements of the manufacturer(s).

Tables 12.3 and 12.4 describe some of the more common tests undertaken during the commissioning phase of an EESS installation.

Note: Not all of these tests will be required on every installation and additional tests may be required on many installations.

Section 12 – EESS inspection and testing

Table 12.3   Equipment and systems tests

Equipment and system tests	Purpose
Data acquisition & control	Tests to ensure that the EESS data acquisition and control systems are performing correctly.
Internet connectivity	Tests to ensure that the EESS is correctly configured and communicating.
Displays/alarms	Checks to ensure correct function of any user displays or alarm mechanisms.
Charge-discharge	Checks that EESS can correctly operate charge and discharge functions.Functional checks of batteries as recommended by the manufacturer or relevant standards (e.g. BS EN 61427-1 recommends battery capacity tests during commissioning).
Thermal management	Checks to ensure correct function of any thermal management systems.
Ventilation	Checks to ensure correct function of any ventilation systems and controls.
Fire systems	Checks to ensure correct function of any fire detection or management systems.
Wet chemical checks	Test to validate correct status of any liquids contained within the EESS (e.g. using hydrometer to check a lead-acid battery).
Ancillary equipment	Checks to ensure correct function of any ancillary equipment forming part of the EESS (e.g. pumps, control valves etc.).

Table 12.4   Application tests

Application tests	Purpose
G98/G99	Tests for EESS that can operate in parallel to the grid ('grid connected'). At the most basic these tests may simply be a 'loss of mains test' to ensure that the system operates correctly when the mains supply is lost. For larger systems, a longer test sequence, such as G99 'injection testing', may be required.
Export limitation (G100)	Where the EESS is part of a system that has been specified to control site export to a specific limit (export from on-site generation or the battery), a suite of tests will be required to prove the correction function of these systems. The commissioning tests expected by a DNO during the commissioning of an export limitation system are described in Engineering Recommendation G100.
Ancillary services	Where a system is providing services to support the grid, specific commissioning tests will be required to prove that the system meets the requirements of that service. These tests typically involve simulating specific grid conditions and then monitoring the EESS to ensure it responds in the correct manner. An example of these tests would be where a test set is used to inject a specific grid frequency change – and then the EESS is monitored to ensure that it has a suitable charge/discharge ramp response.
Revenue metering/monitoring	Where an EESS derives income from providing a service, metering and data acquisition systems are deployed to monitor the EESS and allow revenue to be collected. The correct operation and accuracy of these systems may need to be tested during commissioning. These tests may also be required for other meters in the system, such as those utilized to determine EESS usage as part of any warranty conditions.
Back-up – power quality	Where an EESS provides a back-up or an island mode power supply, the quality of that supply may need to be verified during commissioning. These tests may vary from simple checks of voltage and frequency to a more detailed examination of wave-shape and harmonic content.

Section 12 – EESS inspection and testing

Table 12.4   *Cont.*

Application tests	Purpose
Backup – changeover	Where an EESS provides a back-up power supply, the transition between on-grid and island mode operation may require functional testing. Information on the requirements for systems providing back-up functionality is provided in Sections 8.3 and 9.4. Commissioning tests need to determine the correct function of the measures adopted and may include: • correct function of changeover switches/relays; • correct function of neutral-earth switching; • correct separation of essential and non-essential supplies; and • measurement of earth electrode resistance for the consumer's earth electrode.

Section 13 N

EESS handover and documentation

13.1 System handover

Handover is the process by which operation and maintenance information is provided for a completed system, once installation and an appropriate amount of commissioning is completed. The information should be available in an appropriate form for the intended audience, which may include:

(a) clients/system owners;
(b) users (this may be the client for EESS in dwellings, or operators for commercial or joint ownership systems); and
(c) maintainers and those decommissioning and/or disposing of EESS or its components.

The process should include:

(a) identifying key parts of the system and explaining their function to users and maintainers;
(b) providing health and safety information for residual risks that will be present for operators, maintainers and those involved in future decommissioning and disposal;
(c) providing instructions on:
　(i) normal operation, including how to identify faults or inefficient operations; and
　(ii) maintenance requirements and procedures;
(d) provision of system manuals, drawings, test certificates and similar documentation; and
(e) responding to any client queries.

Overall, compliance with the CDM Regulations is required for system documentation, to facilitate safety in operation and maintenance of the system.

13.2 Minimum system documentation

It is important for the continued efficiency and safety of the EESS that adequate information is provided at handover. Table 13.1 summarizes the type of information that is typically provided.

Table 13.1 EESS documentation – basic data

Item	Notes
System designer	Name, address and contact details.
System installer	Name, address and contact details.
Basic system information	• power rating of EESS and inverters; • total storage capacity of batteries; • make, model and quantity of key components; • installation and commissioning dates; and • client and site details.
Datasheets	Data sheets of key components.

Section 13 – EESS handover and documentation

Table 13.2 Test and commissioning data

Item	Notes
AC system DC system	Inspection and test results and certificates (BS 7671).
Storage batteries	Test and inspection results to demonstrate conformity to the BS EN 62485 series as appropriate.
G98/G99 protection	• test results and certificates; and • settings.

Table 13.3 EESS core components documentation scope

Item	Notes
Wiring/connection diagrams	• DC system schematics; and • detailed line diagrams showing protection, isolation, switching etc.
Charger and inverter	• information on configuration; • information on ventilation and other environmental requirements; • component safe isolation/maintenance procedures; and • overcurrent protection details.
Storage batteries	• battery chemistry; • battery charging, discharging and monitoring requirements; • information on configuration and voltage; • location diagrams; • ventilation and other environmental requirements; • weight; • safe isolation and maintenance/replacement procedures; and • overcurrent protection details.
Earthing and overvoltage protection	• details of earthing and bonding system (where the system operates in island mode as well as on-grid, the earthing system for each mode of operation must be clearly identified); • details of bonding to any lighting protection system; and • details of SPDs.
Physical layout	• physical layout of the EESS and its location on plan drawings; and • details of location of buried or hidden cabling.

Section 13 – EESS handover and documentation

Table 13.4   EESS AC and remaining DC system documentation scope

Item	Notes
Wiring/connection diagrams	• AC system schematics; and • detailed line diagrams showing protection, isolation, switching etc.
AC system	• AC isolators – including location and rating; • overcurrent protection – including location and rating; and • residual current protection – including location and rating.
Earthing and overvoltage protection	• details of the earthing and bonding system (where the system operates in island mode as well as on-grid, the earthing system for each mode of operation must be clearly identified); • details of bonding to any lighting protection system; and • details of SPDs.
G99 protection	• type and location; and • settings.

Table 13.5   Operation and maintenance data

Item	Notes
Manuals	• charger/inverter; • storage batteries; and • monitoring systems/displays.
Warranty	• start date, duration and details
Normal operation	• how to verity that the system is operating normally; • explanation of alarm/messaging systems and indications on key controls or protection; and • user awareness of power limitations in island mode.
Shutdown	• emergency isolation/shutdown procedures.
Fault-finding	• fault-finding procedure; and • service/emergency contact numbers.
Maintenance	• periodic maintenance/cleaning requirements; • service/inspection intervals; and • information on any maintenance contracts.
Disposal	• information on disposal or 'return to supplier' arrangements for components and spares, e.g. batteries, and any environmental and safety constraints for transportation and disposal.

13.3 Larger projects (typically commercial and industrial, or multi-occupancy residential)

For larger projects, the information required is likely to be more extensive, based on the monetary value of the assets being specified and installed, and to help manage safety, operational, and business continuity risks. The level of detail required for operation and maintenance (O&M) information should be specified and delivered in line with the following standards and industry guidance:

(a) BSRIA BG 1/2007 *Handover, O&M Manuals and Project Feedback*;
(b) BSRIA BG 2/2004 *Computer-based Operating and Maintenance Manuals*;
(c) CIBSE *Guide M: Maintenance Engineering and Management*;
(d) BS EN 61082-1; and
(e) BS EN IEC/IEEE 82079-1.

Section 13 – EESS handover and documentation

Current best-practice is for computer aided design (CAD) drawings and electronic O&M manual information to incorporate building information modelling (BIM) data at an appropriate level.

It is recommended that O&M manuals for large EESS are specified to have a Class C or Class D level of detail in accordance with BSRIA BG 1/2007 (see Table 13.6), in order that the operating organization can ensure the continued effectiveness of the EESS and the investment can be optimized, and also that any commitments made in planning and other consents/approvals can be maintained.

Table 13.6   BSRIA level of detail classification for system-based manuals, courtesy BSRIA BG 1/2007

Detail Class	Level of detail required
Class A	Basic record drawings and manufacturer's literature or manuals.
Class B	Record drawings, manufacturer's operating instructions and manuals, test certificates and parts lists.
Class C	Purpose and planning information, technical documentation (including record drawings), full operating instructions for each system, detailed maintenance instructions and schedules, parts lists, modification instructions, and disposal instructions.
Class D	As Class C, with procedures and results of all tests undertaken during the commissioning process and plant and system warranties.

 Section 14 N

EESS operation and maintenance

14.1 General

Maintenance activities generally fall into the following categories:

(a) proactive maintenance – for maintenance activities that can be scheduled based on known parameters and conditions of components, or as a result of safety practices and standards; these will include:
   (i) scheduled maintenance activities (see Section 14.2); and
   (ii) periodic verification (see Section 14.3 and 14.4).
(b) reactive maintenance, where the maintenance activity results from a fault, component failure or change in conditions.

All maintenance activities, and their scheduling, should comply with the O&M manuals provided at handover (see Section 13).

14.2 Scheduled maintenance

The maintenance schedule will typically include the following types of activities:

(a) cleaning of any fans/ventilation in electronic components (such as those within inverters/chargers/battery enclosures);
(b) checking for accumulation of dust and other contaminants on batteries and terminals;
(c) checking fluid levels where appropriate to the type of battery used; and
(d) battery replacement at longer intervals.

Scheduled maintenance may be linked to service contracts, warranty conditions or performance guarantees.

14.3 Periodic verification

Periodic verification is scheduled to ensure that the EESS remains in a safe and satisfactory operational condition. The regime used will be similar to, or a sub-set of, that used for initial system commissioning.

Non-invasive testing or discharge testing of battery systems, in line with manufacturers' and insurers' requirements, should be accommodated.

The interval may be specified by the client on the advice of the company designing and/or installing the EESS or linked to service and maintenance contracts.

As a general rule, the period between verifications should not exceed that recommended for the AC systems to which the EESS is connected.

Instead of the appropriate initial verification certification provided for electrical systems in accordance with BS 7671, an electrical installation condition report would be used to record and present the results of electrical inspections and tests.

Section 14 – EESS operation and maintenance

It is imperative that those carrying out periodic verification in installations with multiple sources of supply are aware of the relative hazards, as inappropriate isolation techniques in a system that is capable of island mode operation can lead to electric shock.

As with initial verification, it should be noted that earth fault loop impedance and prospective fault current measurements may be distorted if measurements are made with local inverters (from EESS, solar PV, and so on) operating in connected mode. Similarly, if these tests are carried out in island mode with small EESS, the tests may not provide an adequate reading before inverter current limit protection operates. See Section 12.2 for the recommended approach for these tests.

14.4 Regular inspections

Regular inspections occur at shorter intervals than periodic verification and are intended to ensure that the system is operating normally. All sites will benefit from a regular inspection, as this may help in the early identification of faults and thus ensure that the system operation does not become inefficient.

14.5 Disposal of replaced components

The Waste Electrical and Electronic Equipment Regulations sets out regulations for the disposal of electrical and electronic equipment. Companies supplying EESS components have to ensure the recycling of damaged and end-of-life components.

Additionally, the Waste Batteries and Accumulators Regulations require those involved in the supply of batteries to ensure provision for recycling at end-of-life or on failure.

14.6 Ensuring continued compatibility of components

With EESS, as with other electrotechnical systems, it is important to ensure that replacement components are compatible in form, fit and function.

However, where storage batteries are concerned, it is possible that apparently minor changes in specification for either batteries and/or chargers can cause irreversible damage to batteries, or cause them to overheat with potentially devastating results such as explosion or fire.

Where batteries include internal protection (such as some lithium-ion types) it is important to ensure the same safety provisions are in place when replacements are procured.

Should a manufacturer cease production of a particular type or model of component, their assistance should be sought in sourcing a suitable equivalent replacement.

Appendix A

Glossary

AC
Alternating current. An electric current that reverses direction at regular intervals. The mains (grid) is an AC supply. The UK mains has a frequency of 50 Hz (the current reverses direction and back again 50 times a second).

AC-coupled system
A system design where the battery is connected on the AC mains (grid) side of renewable generation inverters.

Amp hour
A unit of charge (in a battery) sufficient to deliver one ampere of current to flow for one hour.

Back-up power supply
A source of power designed to provide an emergency supply in the event that the normal, typically mains (grid), supply fails.

Balancing market
Competitive market used to balance discrepancies between the amount of electricity produced and the current demand, in half-hour trading periods.

Battery effective capacity
The usable capacity of a battery – the amount of energy that can be delivered during normal operation, where the DOD is restricted. The effective capacity is less than the nominal capacity.

Battery management system
System that controls the charge and discharge rates and the parameters of batteries.

Battery nominal capacity
Provided by the manufacturer – describes how much energy the battery can nominally deliver from fully charged, under a certain set of conditions. The nominal capacity is more than the effective capacity.

C-rate
Rating used to compare the discharge rates and battery capacities. See Section 5.2.3.

Connected mode
Mode of operation of a prosumer's electrical installation where the installation is connected to the grid. During connected mode, the installation may be *direct feeding* (importing power from the grid) or *reverse feeding* (exporting power to the grid). (*See BS HD 60364-8-2*)

DC
Direct current. An electric current with a constant direction. The output of a battery or PV array is DC.

DC-coupled system
A system design where the battery is connected on the DC (PV array) side of renewable generation inverters.

Depth of discharge (DOD)
Describes how fully a battery has been discharged during a discharge cycle. It is expressed as a percentage of battery capacity, for example 60 %.

Efficiency: battery (charge/discharge) efficiency
A measure of how effective a battery is throughout the full charge-discharge cycle. For example, a battery with an efficiency of 90 % will deliver 90 kWh for every 100 kWh put into it.

Appendix A – Glossary

Efficiency: system (charge/discharge) efficiency

A measure of how effective the complete system is throughout the full charge-discharge cycle. For example, a system with an efficiency of 80 % will deliver 80 kWh for every 100 kWh put into it.

Energy density

The amount of energy stored within a specific mass. A battery with a high energy density is able to deliver more energy than one of the same size but with a lower energy density.

Essential loads

A subset of a building's loads, usually wired into a separate distribution board, that will still get power when a system runs in island mode (non-essential loads get no power).

Grid-forming inverter

PCE that can operate stand-alone in self-commutating mode.

Grid-independent system

An electrical system that is never connected to the grid and operates solely from local generation and/or locally stored energy. Sometimes grid-independent systems are termed 'off-grid', but for consistency that is not used in this Code of Practice, as the term 'island mode' is used to align with BS HD 60364-8-2. (See also **island mode**).

Inverter: multi-function

A device that is able to function as both a renewable generation inverter and a stand-alone inverter.

Inverter: solar PV

The device that provides the interface between a solar PV array and the mains (grid) in a standard grid-connected PV system. A solar PV inverter will not produce any output without a stable frequency reference connected on its AC side.

Inverter: stand-alone

A device that converts DC (for example, from a battery or PV array) into AC 'mains' voltage and frequency.

Island or island mode

An electrical system that is normally connected to the grid, but is operating in a mode where some or all of the installation is isolated from the grid and is operating solely from an EESS. (*See BS HD 60364-8-2*)

Island mode isolator

Device that disconnects the live conductors of the grid supply when the system switches to island mode.

Mains (grid)

The normal (AC) electricity supply that is provided to homes and businesses. In the UK, in accordance with BS EN 50160, the mains has a nominal frequency of 50 Hz and voltage of:

(a) 230 V for single-phase supplies;

(b) 400 V for three-phase supplies with earthed neutral/star-point; and

(c) 230 V for three-phase, 3-wire supplies earthed at one of the phases.

Note: For most installations within the scope of this Code of Practice, the ESQCR limits voltage variations to between +10 % and −6 % at the origin of the installation.

Mains (grid) charging

Charging a battery using power from the mains (grid).

N-E bond relay

Device that establishes a neutral in parts of the installation disconnected from the grid supply in island mode.

Appendix A – Glossary

PCE
Power conversion equipment. A generic term that covers inverters, rectifiers and other converters, or a unit that is a combination of these.

Power cut
A period where the mains (grid) is unavailable.

Prosumer
An entity or party that can be both a producer and a consumer of electrical energy. (*See BS HD 60364-8-2*)

Protective earthing
Earthing of one or more points in a system, installation or equipment for the purposes of safety. (*See BS 7671*)

Protective conductor (PE)
A conductor used for some measures of protection against electric shock and intended for connecting together any of the following parts:

(i) exposed-conductive-parts;
(ii) extraneous-conductive-parts;
(iii) the main earth terminal;
(iv) earth electrode(s); and
(v) the earthed point of the source, or an artificial neutral.

(*See BS 7671*)

Self-discharge
A battery phenomenon whereby the available charge gradually decreases over time.

Solar surplus
When the electricity generation of a solar PV system is greater than the requirements of the installation that it is connected to, the excess is termed 'solar surplus'.

Time shifting
Storing electricity temporarily within a battery for use later in the day (or week).

Uninterruptible power system (UPS)
Combination of convertors, switches and energy storage devices (such as batteries), constituting a power system for maintaining continuity of load power in case of input power failure. (*See BS EN IEC 62040-1*)

Winter mode
This refers to an operating mode for a solar storage system. In winter mode, to protect the battery from a prolonged period of low state of charge, the system is either put to sleep or disconnected from the battery.

Appendix B

References

IET *Code of Practice for Low and Extra Low Voltage Direct Current Power Distribution in Buildings*, 2015.

IET Technical Briefing: *Practical considerations for d.c. installations*, 2016. Free to download from the IET web-site: https://electrical.theiet.org/guidance-codes-of-practice/publications-by-category/building-management-and-maintenance/practical-considerations-for-dc-installations/

IET *Code of Practice for Grid Connected Solar Photovoltaic Systems*, 2015.

IET *Code of Practice for Electric Vehicle Charging Equipment Installations*, 4th Edition, 2020.

IET Guidance Note 6 *Protection Against Overcurrent*, 8th Edition, 2018.

IET Health & Safety Briefing No. 51c *Arc flash protection*, May 2017.

BS 5839-6:2019. *Fire detection and fire alarm systems for buildings. Part 6: Code of practice for the design, installation, commissioning and maintenance of fire detection and fire alarm systems in domestic premises.*

BS 7430:2011+A1:2015. *Code of practice for protective earthing of electrical installations.*

BS 7671:2018+A1:2020. *Requirements for Electrical Installations* (IET Wiring Regulations, 18th Edition).

BS EN 547-1:1996+A1:2008 *Safety of machinery. Human body measurements. Part 1 – Principles for determining the dimensions required for openings for whole body access into machinery.*

BS EN 547-2:1996+A1:2008 *Safety of machinery. Human body measurements. Part 2 – Principles for determining the dimensions required for access openings.*

BS EN 547-3:1996+A1:2008 *Safety of machinery. Human body measurements. Part 3 – Anthropometric data.*

BS EN 50171:2001. *Central power supply systems.*

BS EN 50160:2010+A3:2019. *Voltage characteristics of electricity supplied by public electricity networks.*

BS EN 50310:2016+A1:2020. *Application of equipotential bonding and earthing in buildings with information technology equipment.*

BS EN 50549-1:2019. *Requirements for generating plants to be connected in parallel with distribution networks. Connection to a LV distribution network. Generating plants up to and including Type B.*

BS EN 60445:2017. *Basic and safety principles for man-machine interface, marking and identification. Identification of equipment terminals, conductor terminations and conductors.*

BS EN 60898-1:2019. *Electrical accessories. Circuit-breakers for overcurrent protection for household and similar installations. Circuit-breakers for a.c. operation.*

BS EN 60898-2:2006. *Electrical accessories. Circuit-breakers for overcurrent protection for household and similar installations. Circuit-breakers for a.c and d.c. operation.*

BS IEC 60898-3:2019. *Electrical accessories. Circuit-breakers for overcurrent protection for household and similar installations. Circuit-breakers for DC operation.*

Appendix B – References

BS EN 61008-1:2012+A12:2017. Residual current operated circuit-breakers without integral overcurrent protection for household and similar uses (RCCBs). General rules.

BS EN 61008-2-1:1995. *Specification for residual current operated circuit-breakers without integral overcurrent protection for household and similar uses (RCCBs). Applicability of the general rules to RCCBs functionally independent of line voltage.*

BS EN 61009-1:2012+A12:2016. *Residual current operated circuit-breakers with integral overcurrent protection for household and similar uses (RCBOs). General rules.*

BS EN 61009-2-1:1995. *Specification for residual current operated circuit-breakers with integral overcurrent protection for household and similar uses (RCBOs). Applicability of the general rules to RCBOs functionally independent of line voltage.*

BS EN 61010-031:2015. *Safety requirements for electrical equipment for measurement, control and laboratory use. Safety requirements for hand-held probe assemblies for electrical measurement and test.*

BS EN 61508-1:2010. *Functional safety of electrical/electronic/ programmable electronic safety-related systems. General requirements.*

BS EN 61508-2:2010. *Functional safety of electrical/electronic/ programmable electronic safety-related systems. Requirements for electrical/electronic/ programmable electronic safety-related systems.*

BS EN 61508-3:2010. *Functional safety of electrical/electronic/ programmable electronic safety-related systems. Software requirements.*

BS EN 61508-4:2010. *Functional safety of electrical/electronic/ programmable electronic safety related systems. Definitions and abbreviations.*

BS EN 61508-5:2010. *Functional safety of electrical/electronic/ programmable electronic safety related systems. Examples of methods for the determination of safety integrity levels.*

BS EN 61508-6:2002. *Functional safety of electrical/ electronic/ programmable electronic safety-related systems. Guidelines on the application of IEC 61508-2 and IEC 61508-3.*

BS EN 61508-7:2010. *Functional safety of electrical/electronic/ programmable electronic safety related systems. Overview of techniques and measures.*

BS EN 62606:2013+A1:2017. *General requirements for arc fault detection devices.*

BS EN IEC 62040-1:2019. *Uninterruptible power systems (UPS). General and safety requirements for UPS.*

BS EN IEC 62040-2:2018. *Uninterruptible power systems (UPS). Electromagnetic compatibility (EMC) requirements.*

BS EN 62040-3:2011. *Uninterruptible power systems (UPS). Method of specifying the performance and test requirements.*

BS EN 62040-4:2013. *Uninterruptible power systems (UPS). Environmental aspects. Requirements and reporting.*

BS EN 61082-1:2015. *Preparation of documents used in electrotechnology. Rules.*

BS EN 62305-1:2011. *Protection against lightning. General principles.*

Appendix B – References

BS EN 62305-2:2012. *Protection against lightning. Risk management.*

BS EN 62305-3:2011. *Protection against lightning. Physical damage to structures and life hazard.*

BS EN 62305-4:2011. *Protection against lightning. Electrical and electronic systems within structures.*

BS EN 62368-1:2014+A11:2017. *Audio/video, information and communication technology equipment. Safety requirements.*

BS EN 62446-1:+A1:2018. *Photovoltaic (PV) systems – Requirements for testing, documentation and maintenance – Part 1: Grid connected systems – Documentation, commissioning tests and inspection.*

BS EN IEC/IEEE 82079-1:2020. *Preparation of instructions for use. Structuring, content and presentation. General principles and detailed requirements.*

BS EN IEC 62485-1:2018. *Safety requirements for secondary batteries and battery installations. General safety information.*

BS EN IEC 62485-2:2018. *Safety requirements for secondary batteries and battery installations. Stationary batteries.*

BS EN IEC 62933-1:2018. *Electrical Energy Storage (EES) systems. Terminology.*

BS EN IEC 62933-2-1:2018. *Electrical Energy Storage (EES) systems. Unit parameters and testing methods.*

BS EN IEC 62933-5-2:2020. *Electrical energy storage (EES) systems. Safety requirements for grid-integrated EES systems. Electrochemical-based systems.*

PD IEC/TS 62933-4-1:2017. *Electrical energy storage (EES) systems. Guidance on environmental issues. General specification.*

BS EN ISO 7010:2020. *Graphical symbols. Safety colours and safety signs. Registered safety signs.*

BS HD 60364-8-2:2011+A11:2019. *Low voltage electrical installations. Prosumer's low voltage electrical installations.*

PD IEC/TR 62368-2:2019. *Audio/video, information and communication technology equipment. Explanatory information related to IEC 62368-1:2018.*

AS/NZS 5139:2019. *Electrical installations – Safety of battery systems for use with power conversion equipment.*

BEAMA The RCD Handbook. *BEAMA Guide to the Selection and Application of Residual Current Devices (RCDs), October 2019.*

BEAMA Guide *Consumer Access Devices. Applications for Data in the Consumer Home Area Network (CHAN) and Wider Market Considerations, 2014.*

BSRIA BG 1/2007 *Handover, O&M Manuals and Project Feedback.*

BSRIA BG 2/2004 *Computer-based Operating and Maintenance Manuals.*

CIBSE Guide M: *Maintenance Engineering and Management.*

Appendix B – References

ENA Engineering Recommendation G98 – Issue 1, Amendment 4 (2019) – *Requirements for the connection of Fully Type Tested Micro-generators (up to and including 16 A per phase) in parallel with public Low Voltage Distribution Networks on or after 27 April 2019.*

ENA Engineering Recommendation G99 – Issue 1, Amendment 5 (2019) – *Requirements for the connection of equipment in parallel with public distribution networks on or after 27 April 2019.*

ENA Engineering Recommendation G100 – Issue 1, Amendment 2 (2018) – *Technical Requirements for Customer Export Limiting Schemes.*

Ammerman, R. F., Gammon, T., Sen, P. K. & Nelson, J. P. *DC-Arc Models and Incident-Energy Calculations.* IEEE Transactions on Industry Applications, Vol. 46, No. 5 (September/October 2010).

Building Regulations 2010 as amended to 2016.

Building (Scotland) Regulations 2004, as amended

Construction (Design and Management) Regulations 2015.

Electrical Equipment (Safety) Regulations 2016.

Electricity at Work Regulations 1989.

Electricity Safety, Quality and Continuity Regulations 2002, as amended.

Electromagnetic Compatibility Regulations 2016.

The Health and Safety (Signs and Signals) Regulations 1996.

Waste Batteries and Accumulators Regulations 2009.

HSE Guidance Note GS38 (Fourth edition): *Electrical test equipment for use on low voltage electrical systems.*

HSE Leaflet INDG 139: *Using electric storage batteries.*

HSE Guidance HSG85: *Electricity at work: Safe working practices.*

 **Appendix C**

System labels and safety signs

C1 General

This Appendix discusses additional signage that may be required when an EESS is introduced into an installation.

All signage should comply as appropriate with the Health & Safety (Signs and Signals) Regulations and BS EN ISO 7010. Other signage may be required for the installation, for example, as required by BS 7671 or recommended by the IET *Code of Practice for Grid-Connected Solar PV*, or as a result of a particular risk assessment.

C2 Battery hazard signage

Suitable hazard warning signs need to be displayed to highlight battery hazards. The appropriate signs will depend on the battery type. For example, for a lead-acid battery, the following signage may be appropriate:

Figure C.1 Example signage for a lead-acid battery enclosure or room

In many circumstances, particularly for larger battery installations, a suitable warning sign at the main incomer (and other generators, where fitted) should also be considered to warn the emergency services that a battery bank is installed on the premises. A suitable sign is shown below:

Appendix C – System labels and safety signs

Figure C.2   Example battery warning label at main incoming switchgear

A warning of the battery voltage should be provided on enclosures of batteries where the battery voltage exceeds 60 V DC.

Figure C.3   Example battery voltage warning label for a 90 V battery

Note: The actual nominal battery voltage should be stated.

Where DC arc flash hazards have been identified by risk assessment (see Appendix E), suitable warning signage will be required, to identify the hazard and identify safe working practices. An example is provided in Figure C.4.

Figure C.4   Example of additional signage for rooms or enclosures where risk of DC arc flash exists

Note: Example of wording only. If this wording is used, the actual assessed arc flash hazard category for PPE would be stated (see Appendix E).

C3   Signage for multiple supplies

Signs should also reference all the sources of generation installed on the site in accordance with BS 7671 and as required by ENA Engineering Recommendations G98 and G99. An example of the 'dual supply' sign for use on sites with multiple sources of generation is shown below:

Figure C.5   G98/G99 warning label for multiple supplies (image courtesy of ENA)

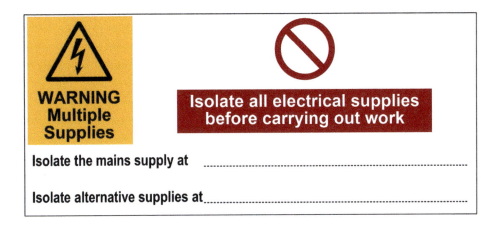

Appendix D

Standards

Standards are available for practitioners to purchase, view or borrow from trade organizations such as the IET and the ECA and from public libraries.

D1   Normative references

These standards are generally required in specifying or installing EESS in accordance with this Code of Practice.

Note: Not all standards listed here will be relevant to every installation.

D1.1   British and European standards for installation

Standard	Commentary on applicability for EESS
BS 7430:2011+A1:2015. *Code of practice for protective earthing of electrical installations*	BS 7430 is a key standard referenced from BS 7671 for the protective earthing of the LV electrical installation. It is relevant to the provision of additional earthing for the EESS to operate in island mode and for grid-independent systems that require earthing.
BS 7671:2018+A1:2020. *Requirements for Electrical Installations* (IET Wiring Regulations, 18th Edition)	Standard for the safety of LV electrical installations in the UK that do not form part of the electricity distribution system, nor certain other exempted installations and equipment. Applicable to the installation design, initial verification, safe maintenance, and periodic inspection and testing of both LV EESS installations and LV electrical installations as a whole.
BS EN 50549-1:2019. *Requirements for generating plants to be connected in parallel with distribution networks. Connection to a LV distribution network. Generating plants up to and including Type B*	See Section 10.2.
BS EN IEC 62933-1:2018. *Electrical Energy Storage (EES) systems. Terminology*	The BS EN IEC 62933-series of standards is currently in development, and relates to grid-connected EESS.
BS EN IEC 62933-2-1:2018. *Electrical Energy Storage (EES) systems. Unit parameters and testing methods*	
BS EN IEC 62933-5-2:2020. *Electrical energy storage (EES) systems. Safety requirements for grid-integrated EES systems. Electrochemical-based systems*	
PD IEC/TS 62933-4-1:2017. *Electrical energy storage (EES) systems. Guidance on environmental issues. General specification*	
BS EN IEC 62485-1. *Central power supply systems*	This standard covers central power systems used for an independent energy supply to essential safety equipment, typically emergency lighting systems. This standard covers systems that are permanently connected to AC supplies not exceeding 1,000 V and that use batteries as the alternative power source. Power systems for fire alarm systems are excluded.
BS EN IEC 62485-2. *Safety requirements for secondary batteries and battery installations. General safety information*	The BS EN IEC 62485 series of standards covers secondary (rechargeable) battery installations for systems where no specific product standard exists. It is relevant to the safe design, installation and maintenance of the batteries, and their charging equipment, their housings and the facilities in which they will be installed. Currently, the series of standards is primarily aimed at NiCd and lead-acid battery types.

Appendix D – Standards

Standard	Commentary on applicability for EESS
BS EN 62305-1:2011. *Protection against lightning. General principles*	BS EN 62305 series of standards, covering lightning protection, is relevant to designers of electrical installations, when ensuring that adequate surge protection is in place. The means of earthing, and hence lightning protection provisions, may change when an EESS is operating in island mode. The impact of external services to the installation, such as copper telecommunications, should be considered.
BS EN 62305-2:2012. *Protection against lightning. Risk management*	
BS EN 62305-3:2011. *Protection against lightning. Physical damage to structures and life hazard*	
BS EN 62305-4:2011. *Protection against lightning. Electrical and electronic systems within structures*	
BS EN ISO 7010:2012+A5:2015. *Graphical symbols – Safety colours and safety signs – Registered safety signs (ISO 7010:2011)*	Standard for the presentation and use of safety signage.
BS HD 60364-8-2:2011+A11:2019. *Low voltage electrical installations. Prosumer's low voltage electrical installations.*	This standard, which is intended to accompany and be read in conjunction with BS 7671, provides functional requirements relating to the design and implementation of prosuming electrical installations. It is currently being considered for inclusion in BS 7671 as Chapter 82.

D1.2 British and European standards for EESS converters and batteries

Note 1: This section focuses on standards required specifically for EESS components, and not the general standards for electrical accessories, protective devices, switchgear and controlgear assemblies, and similar items, that are required for electrical installations in general.

Note 2: Whilst battery standards are listed, it is acknowledged that not all of the chemistries in current use are represented by existing standards, and that some of the types listed have limitations in EESS.

Standard
BS EN 50549-1:2019. *Requirements for generating plants to be connected in parallel with distribution networks. Connection to a LV distribution network. Generating plants up to and including Type B*
BS EN 60146-1-1:2010. *Semiconductor converters. General requirements and line commutated converters. Specification of basic requirements*
PD IEC TR 60146-1-2:2019. *Semiconductor converters. General requirements and line commutated converters. Application guide*
BS EN 60146-1-3:1993. *Semiconductor convertors. General requirements and line commutated convertors. Transformers and reactors*
BS EN 60146-2:2000. *Semiconductor convertors. General requirements and line commutated convertors. Self-commutated semiconductor converters including direct d.c. converters*
BS EN 60622:2003. *Secondary cells and batteries containing alkaline or other non-acid electrolytes. Sealed nickel-cadmium prismatic rechargeable single cells*
BS EN 60896-11:2003. *Stationary lead-acid batteries. General requirements and methods of test. Vented types. General requirements and methods of tests*
BS EN 60896-21: 2004. *Stationary lead-acid batteries. Valve regulated types. Methods of test*
BS EN 60896-22:2004. *Stationary lead-acid batteries. Valve regulated types. Requirements*
BS EN 61056-1:2012. *General purpose lead-acid batteries (valve-regulated types). General requirements, functional characteristics. Methods of test*
BS EN 61056-2:2012. *General purpose lead-acid batteries (valve-regulated types). Dimensions, terminals and marking*
BS EN 61427-1:2013. *Secondary cells and batteries for renewable energy storage. General requirements and methods of test. Photovoltaic off-grid application*
BS EN 61427-2:2017. *Secondary cells and batteries for renewable energy storage. General requirements and methods of test. On-grid applications*
BS EN 61429:1997, IEC 61429:2015. *Marking of secondary cells and batteries with the international recycling symbol ISO 7000-1135 and indications regarding directives 93/86/EEC and 91/157/EEC*
BS EN 62109-1:2010. *Safety of power converters for use in photovoltaic power systems. General requirements*

Appendix D – Standards

Standard
Safety of power converters for use in photovoltaic power systems. General requirements *Safety of power converters for use in photovoltaic power systems. Particular requirements for inverters*
BS EN 62116:2014. *Utility-interconnected photovoltaic inverters. Test procedure of islanding prevention measures*
BS EN 62477-1:2012+A1:2017. *Safety requirements for power electronic converter systems and equipment. General*
BS EN 62619:2017. *Secondary cells and batteries containing alkaline or other non-acid electrolytes. Safety requirements for secondary lithium cells and batteries, for use in industrial applications*
BS EN 62620:2015. *Secondary cells and batteries containing alkaline or other non-acid electrolytes. Secondary lithium cells and batteries for use in industrial applications*
BS EN 62675:2014. *Secondary cells and batteries containing alkaline or other non-acid electrolytes. Sealed nickel-metal hydride prismatic rechargeable single cells*
BS EN IEC 62909-1:2018. *Bi-directional grid connected power converters. General requirements*
BS EN IEC 62909-2:2019. *Bi-directional grid-connected power converters. Interface of GCPC and distributed energy resources*

D1.3 Industry standards

Standard	Notes
ENA Engineering Recommendation G99 – *Recommendations for the Connection Of Generating Plant to The Distribution Systems of Licensed Distribution Network Operators*	Distribution network operator industry requirements for installations containing grid-connected EESS.
ENA Engineering Recommendation G98 – *Recommendations for the Connection of Type Tested Small-scale Embedded Generators (Up to 16A per Phase)*	
ENA Engineering Recommendation G100 – Issue 1 (2016) – *Technical Requirements for Customer Export Limiting Schemes*	
Construction Industry Research and Information Association *CDM 2015 – Construction work sector guidance for designers, fourth edition (C755)*	Guidance to help designers meet their CDM Regulations duties for construction risks.
Construction Industry Research and Information Association *CDM 2015 – Workplace 'in-use' guidance for designers, second edition (C756)*	Guidance to help designers meet their CDM Regulations duties for in-use risks for workplaces.

Appendix D – Standards

D2   Related standards

These standards may also be useful to those designing, installing and maintaining EESS. The list is not intended to be exhaustive.

D2.1   British and European standards

Standard	Notes
BS EN 50171:2001. *Central power supply systems*	This standard covers central power systems used for an independent energy supply to essential safety equipment, typically emergency lighting systems. This standard covers systems that are permanently connected to AC supplies not exceeding 1,000 V and that use batteries as the alternative power source. Power systems for fire alarm systems are excluded.
BS EN 60079-10-1:2015. *Explosive atmospheres. Classification of areas. Explosive gas atmospheres*	The BS EN 60079-series of standards is applicable to equipment for use in explosive atmospheres.
BS EN 60079-14:2014. *Explosive atmospheres. Electrical installations design, selection and erection*	
BS EN 62040-1:2019. *Uninterruptible power systems (UPS). General and safety requirements for UPS*	The BS EN 62040 series of standards covers the safety, electromagnetic compatibility and environmental aspects of UPS other than central power supply systems, which comply with BS EN 50171.
BS EN 62040-2:2018. *Uninterruptible power systems (UPS). Electromagnetic compatibility (EMC) requirements*	
BS EN 62040-3:2011. *Uninterruptible power systems (UPS). Method of specifying the performance and test requirements*	
BS EN 62040-4:2013. *Uninterruptible power systems (UPS). Environmental aspects. Requirements and reporting*	
BS EN 62040-5-3:2017. *Uninterruptible power systems (UPS). DC output UPS. Performance and test requirements*	
BS EN 62368-1:2014+A11:2017. *Audio/video, information and communication technology equipment. Safety requirements*	Standard for the safety of electronic equipment for applications such as audio, video, information and communication technology. It replaces a number of older product standards for this family of equipment, including BS EN 60950-1.

D2.2   International standards

Technical Committee TC120 of the IEC is currently addressing industry standards in the area of storage systems. These are draft at the time of publication, and have therefore not been referenced.

D2.3   National standards from other countries

Standard	Notes
AS/NZS 5139:2019. *Electrical installations – Safety of battery systems for use with power conversion equipment.*	This standard contains a published methodology for DC arc flash risk assessment, which is outlined in Appendix E.

≡ Appendix E

DC arc flash risk assessment

E1   Introduction

This Appendix provides guidance on arc flash risk assessment. Two cases are considered:

(a) pre-assembled battery systems where a manufacturer has conducted a thorough analysis of the risks, and has provided comprehensive installation and maintenance instructions (see Section E2); and
(b) other types of battery systems, where the designer and/or installer must conduct a risk assessment for themselves (see Section E3).

E2   Pre-assembled battery systems

Figure E.1   Approach to DC arc flash hazard assessment for pre-assembled battery systems

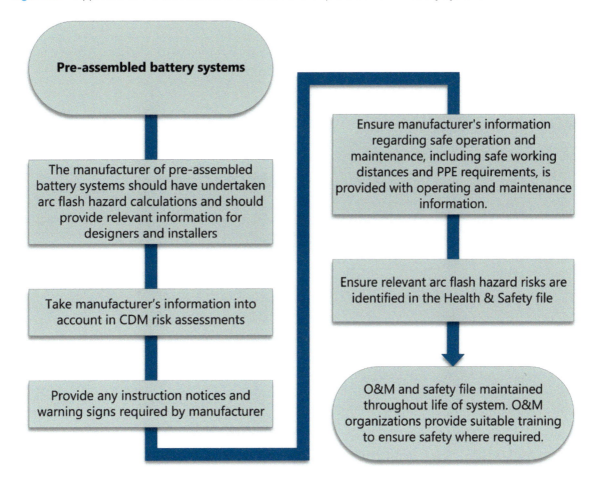

Appendix E – DC arc flash risk assessment

E3   Approach for other battery systems

E3.1   Introduction

The approach, outlined in Figure E.2, with guidance in E3.2 to E3.5, is based on the methodology in the Australia/New Zealand standard AS/NZS 5139:2019 *Electrical installations – Safety of battery systems for use with power conversion equipment* with kind permission of Standards Australia Limited. Copyright in AS/NZS 5139:2019 vests jointly in Standards Sustralia Limited and The Sovereign in Right of New Zealand acting by and through the Ministry of Business, Innovation and Employment (MBIE).

It is suitable for battery system voltages up to 1,000 V DC.

Alternative approaches can be used to calculate arc flash incident energy and arc flash boundary. It is recommended that designers choose a methodology that has been subject to industry peer review, such as the IEEE paper *DC-Arc Models and Incident-Energy Calculations* (by Ammerman, Gammon, Sen and Nelson, IEEE Transactions on Industry Applications, Vol. 46, No. 5 (September/October 2010)).

Note 1: Incident energy and ratings of arc flash PPE in the relevant standards are expressed in calories per square centimetre (cal/cm^2), as conversion from calories to joules is not linear with temperature.

Note 2: In some systems, charging current contributes little to DC arc fault currents, because it is limited by hard current limits in PCE. Where this is not the case, for example, where direct rectified AC protected by overcurrent protective devices such as fuses or circuit-breakers is used to charge batteries, the contribution of the charging supply to the DC arc fault current should also be taken into account.

Figure E.2   Approach to DC arc flash hazard assessment for battery systems that are not pre-assembled

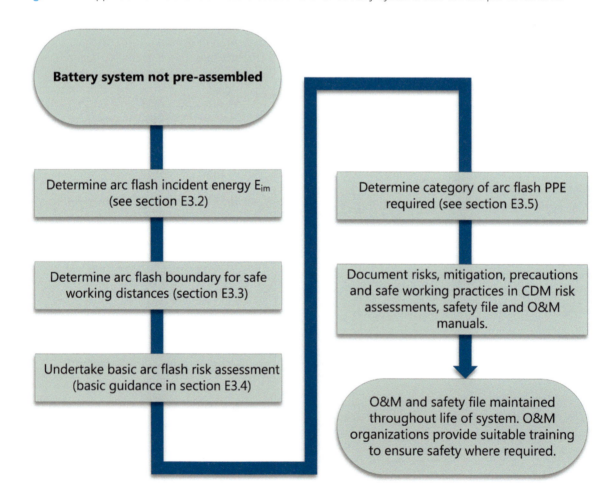

Appendix E – DC arc flash risk assessment

E3.2   Calculating arc flash incident energy Reproduced with the permission of Standards Australia Limited. Copyright in AS/NZS 5139:2019 vests jointly in Standards Australia Limited and The Sovereign in Right of New Zealand acting by and through the Ministry of Business, Innovation and Employment (MBIE).

Arc flash incident energy is used to determine the selection of the relevant PPE required to work in the vicinity of the area at risk of arc flash, and also to inform design, fire safety and safe working practice risk assessments of the level of hazard involved.

The arc flash incident energy E_{im} is calculated as follows:

$$E_{im} = \frac{U_{sys} \times I_{arc} \times T_{arc} \times k_{mf}}{100 \times D^2} \quad \text{where:}$$

E_{im}	is the estimated DC arc flash incident energy at the maximum power point, in cal/cm^2.
U_{sys}	is the battery system voltage at the point of fault, in V.
I_{arc}	is the arcing current, in A, equivalent to half the battery prospective fault current in A. Where appropriate, the contribution of the prospective fault current of the charging supply should be taken into account (see Note 2 in Section E3.1).
T_{arc}	is the arcing time, in seconds:

    (a)   where there is no string fusing, assume the arcing time is 2 s; and
    (b)   where string overcurrent protection is provided, then either:
        (i)   the arcing time graphs for the protective device should be consulted; or
        (ii) where it is known that the arcing time will be less than 0.1 s, use $T_{arc} = 0.1$.

D	is the working distance, in cm. A maximum distance of 45 cm should be used.
k_{mf}	is a multiplying factor:

    (a)   for batteries installed in battery rooms or containers, $k_{mf} = 1.5$; and
    (b)   for batteries installed in battery enclosures, $k_{mf} = 3$.

Figures E.3, E.4 and E.5 show examples of parameters to use for various examples of faults. It should be noted that these are generalizations: the designer should assess the actual arrangement of particular batteries.

Appendix E – DC arc flash risk assessment

Figure E.3 Examples of faults with no protection (adapted from Figure F.1 of AS/NZS 5139 with the permission of Standards Australia Limited. Copyright in AS/NZS 5139:2019 vests jointly in Standards Australia Limited and The Sovereign in Right of New Zealand acting by and through the Ministry of Business, Innovation and Employment (MBIE)).

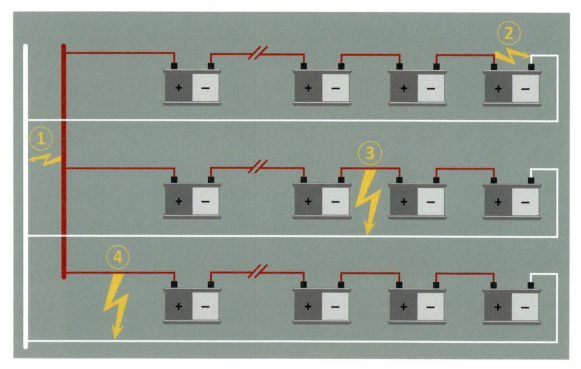

1 DC bus fault:

- U_{sys} = DC bus voltage
- I_{arc} = sum of string arcing fault currents (see Note)
- T_{arc} = 2 s

2 Single cell/module fault:

- U_{sys} = cell/module voltage
- I_{arc} = cell/module arcing fault current
- T_{arc} = 2 s

3 Multiple cell/module fault:

- U_{sys} = sum of voltages of cells/modules in string to point of fault
- I_{arc} = sum of cell/module arcing fault currents in string to point of fault
- T_{arc} = 2 s

4 String fault:

- U_{sys} = DC bus voltage
- I_{arc} = sum of string arcing fault currents (see Note)
- T_{arc} = 2 s

Note: Where the prospective fault current of the charging supply exceeds 10 % of the string prospective fault current, this should also be taken into account (see Note 2 in Section E3.1).

Appendix E – DC arc flash risk assessment

Figure E.4 Examples of faults with string protection (adapted from Figure F.2 of AS/NZS 5139 with the permission of Standards Australia Limited. Copyright in AS/NZS 5139:2019 vests jointly in Standards Australia Limited and The Sovereign in Right of New Zealand acting by and through the Ministry of Business, Innovation and Employment (MBIE)).

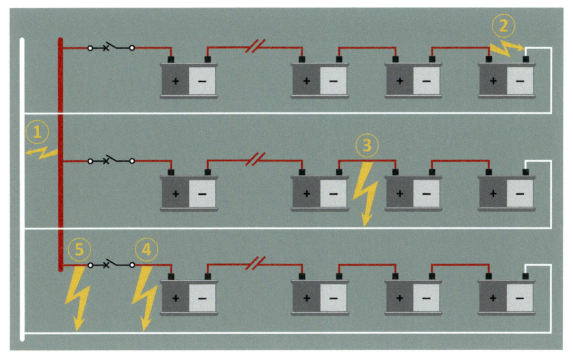

1   DC bus fault:

- U_{sys} = DC bus voltage
- I_{arc} = sum of string arcing fault currents (see Note)
- T_{arc} = 0.1 s or as per arcing time graph of protective devices

2   Single cell/module fault:

- U_{sys} = cell/module voltage
- I_{arc} = cell/module arcing fault current
- T_{arc} = 2 s

3   Multiple cell/module fault:

- U_{sys} = sum of voltages of cells/modules in string to point of fault
- I_{arc} = sum of cell/module arcing fault currents in string to point of fault
- T_{arc} = 2 s

4   String fault on unprotected side of string protective device:

- U_{sys} = DC bus voltage
- I_{arc} = string arcing fault current (see Note)
- T_{arc} = 2 s

5   String fault on bus side of string protective device:

- U_{sys} = DC bus voltage
- I_{arc} = sum of string arcing fault currents (see Note)
- T_{arc} = 0.1 s or as per arcing time graph of protective devices

Note: Where the prospective fault current of the charging supply exceeds 10 % of the string prospective fault current, this should also be taken into account (see Note 2 in Section E3.1).

Appendix E – DC arc flash risk assessment

Figure E.5 Examples of faults with inter-string protection (adapted from Figure F.3 of AS/NZS 5139 with the permission of Standards Australia Limited. Copyright in AS/NZS 5139:2019 vests jointly in Standards Australia Limited and The Sovereign in Right of New Zealand acting by and through the Ministry of Business, Innovation and Employment (MBIE)).

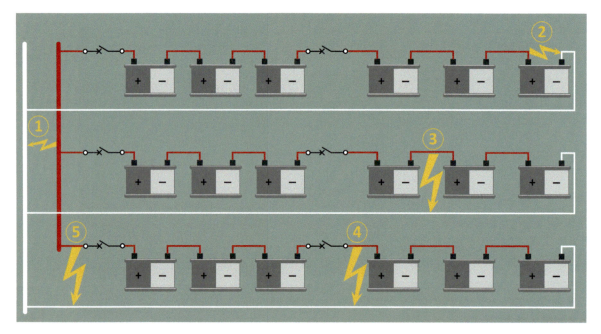

1 DC bus fault:

- U_{sys} = DC bus voltage
- I_{arc} = sum of string arcing fault currents (see Note)
- T_{arc} = 0.1 s or as per arcing time graph of protective devices

2 Single cell/module fault:

- U_{sys} = cell/module voltage
- I_{arc} = cell/module arcing fault current
- T_{arc} = 2 s

3 Multiple cell/module fault:

- U_{sys} = sum of voltages of cells/modules in string to point of fault
- I_{arc} = sum of cell/module arcing fault currents in string to point of fault
- T_{arc} = 2 s

4 String fault on unprotected side of inter-string protective device:

- U_{sys} = DC bus voltage/2
- I_{arc} = string arcing fault current
- T_{arc} = 2 s

5 String fault on bus side of string protective device:

- U_{sys} = DC bus voltage
- I_{arc} = sum of string arcing fault currents (see Note)
- T_{arc} = 0.1 s or as per arcing time graph of protective devices

Note: Where the prospective fault current of the charging supply exceeds 10 % of the string prospective fault current, this should also be taken into account (see Note 2 in Section E3.1).

Appendix E – DC arc flash risk assessment

E3.3 Arc flash boundary Reproduced with the permission of Standards Australia Limited. Copyright in AS/NZS 5139:2019 vests jointly in Standards Australia Limited and The Sovereign in Right of New Zealand acting by and through the Ministry of Business, Innovation and Employment (MBIE).

The arc flash boundary is the distance from electrical equipment at which the arc flash incident energy is equal to 1.2 cal/cm^2. PPE of an appropriate category (see Section E3.5) should be worn within the arc flash boundary.

The arc flash boundary is calculated as follows:

$$D_{afb} = \sqrt{\frac{U_{sys} \times I_{arc} \times T_{arc} \times k_{mf}}{120}} \quad \text{where:}$$

D_{afb}	is the arc flash boundary distance from the electrical equipment, in cm.
U_{sys}	is the battery system voltage, in V.
I_{arc}	is the arcing current, in A, equivalent to half the battery prospective fault current in A. Where appropriate, the contribution of the prospective fault current of the charging supply should be taken into account (see Note 2 in Section Section E3.1).
T_{arc}	is the arcing time, in seconds:

    (a) where there is no string fusing, assume the arcing time is 2 s; and

    (b) where string overcurrent protection is provided, then either:

        (i) the arcing time graphs for the protective device should be consulted; or

        (ii) where it is known that the arcing time will be less than 0.1 s, use T_{arc} = 0.1.

k_{mf} is a multiplying factor:

    (a) for batteries installed in battery rooms or containers, k_{mf} = 1.5; and

    (b) for batteries installed in battery enclosures, k_{mf} = 3.

Figures E.3, E.4 and E.5 show examples of parameters to use for various examples of faults. It should be noted that these are generalizations: the designer should assess the actual arrangement of particular batteries.

E3.4 Arc flash risk assessment

Once the incident energy and boundary are known, the designer should conduct an appropriate risk assessment based on the intended installation method, and considering any appropriate electrical, mechanical and fire protection in place.

The risk assessment should also consider safety, and, where necessary, safe working practices, for installation, operation, maintenance and decommissioning of the system. Whilst a suitable and sufficient risk assessment is beyond the scope of this Appendix, and will vary from system to system, Table E.1 outlines some example overall system design and installation considerations based on incident energy levels.

Appendix E – DC arc flash risk assessment

Table E.1 Example system design and installation considerations for arc flash risk assessment

Calculated E_{im} (cal/cm²)	Consequence description	System design and installation considerations (basic example)
$0.0 \leq E_{im} < 1.2$	Insignificant	n/a
$1.2 \leq E_{im} < 4.0$	Minor	In dwellings, recommend access is restricted by the use of a tool.
$4.0 \leq E_{im} < 8.0$	Moderate	As Minor, plus: **(a)** additional fire precautions are recommended if installed within or adjacent to buildings with habitable rooms; and **(b)** in dwellings, recommend that the battery can be isolated into blocks with $E_{im} < 4.0$ cal/cm².
$8.0 \leq E_{im} < 40$	Major	As Moderate, plus: **(a)** not recommended for installations associated with dwellings; and **(b)** recommend that the battery can be isolated into blocks with $E_{im} < 8.0$ cal/cm².
$E_{im} \geq 40$	Catastrophic	As Major, plus: **(a)** not recommended for installation within any building. **(b)** separate external battery container may be required. Special working procedures and precautions are likely to be necessary.

E3.5 Selection of appropriate grade of personal protective equipment (PPE)

Example categories are provided in Table E.2.

Table E.2 Example of selection of PPE based on calculated incident energy (adapted from IET Health & Safety Briefing No. 51c *Arc flash protection*, May 2017)

Calculated E_{im} (cal/cm²)	Consequence description	PPE category	PPE details (basic examples)
$0.0 \leq E_{im} < 1.2$	Insignificant	0	1-layer untreated cotton (covering all body); polycarbonate safety spectacles; lightweight cotton gloves.
$1.2 \leq E_{im} < 4.0$	Minor	1	Cotton undergarments; 1-layer flame-retardant (FR) workwear; safety helmet; polycarbonate safety spectacles; lightweight FR gloves.
$4.0 \leq E_{im} < 8.0$	Moderate	2	As PPE Category 1, but with 2-layer FR outer workwear with wrist closures, and a full-face polycarbonate visor. An FR single-layer balaclava may also be worn to protect the face.
$8.0 \leq E_{im} < 40$	Major	3	3-layer FR outer workwear with cotton undergarments and FR shirt; a full-face hood or visor with safety spectacles underneath; chrome leather gauntlets.
$E_{im} \geq 40$	Catastrophic	4	Typically 4-layer FR outer workwear; FR and electrically insulated footwear and suitable FR material spats to close off the ankle area; FR gloves or chrome leather gauntlets; a hood constructed from a triple-layer FR material, with a sewn-in polycarbonate face-shield with a minimum of two panels of suitable thickness, one coated with a gold film for ultra-violet (UV) protection.

Appendix E – DC arc flash risk assessment

E3.6 Examples of arc flash incident energy and arc flash protection boundary distances

Tables E.3 to E.6 demonstrate that use of appropriate inter-string and battery system terminal protection can significantly influence the level of incident energy potential from an arc event. These tables assume the charging current is provided by PCE with appropriate current limit, and therefore the prospective arc fault current contribution of the charging supply is negligible (see Note 2 in Section E3.1).

Table E.3 Arc flash incident energy and arc flash protection boundary distances for 24 V battery systems (Table F.1 in AS/NZS 5139)

Inter-string protection	DC arc voltage	Bolted fault current at point of activity	Arcing time	Multiplying factor	Incident energy	Arc flash boundary
	V	kA	s		cal/cm²	cm
No	24	3	2	3	1.067	42.43
	24	6	2	3	2.133	60.00
	24	9	2	3	3.200	73.48
	24	12	2	3	4.267	84.85
	24	15	2	3	5.333	94.87
	24	18	2	3	6.400	103.92
	24	21	2	3	7.467	112.25
	24	24	2	3	8.533	120.00
	24	27	2	3	9.600	127.28
	24	30	2	3	10.667	134.16
Yes	24	3	0.1	3	0.053	9.49
	24	6	0.1	3	0.107	13.42
	24	9	0.1	3	0.160	16.43
	24	12	0.1	3	0.213	18.97
	24	15	0.1	3	0.267	21.21
	24	18	0.1	3	0.320	23.24
	24	21	0.1	3	0.373	25.10
	24	24	0.1	3	0.427	26.83
	24	27	0.1	3	0.480	28.46
	24	30	0.1	3	0.533	30.00

Reproduced with the permission of Standards Australia Limited. Copyright in AS/NZS 5139:2019 vests jointly in Standards Australia Limited and The Sovereign in Right of New Zealand acting by and through the Ministry of Business, Innovation and Employment (MBIE).

Appendix E – DC arc flash risk assessment

Table E.4 Arc flash incident energy and arc flash protection boundary distances for 48 V battery systems (Table F.2 in AS/NZS 5139)

Inter-string protection	DC arc voltage	Bolted fault current at point of activity	Arcing time	Multiplying factor	Incident energy	Arc flash boundary
	V	kA	s		cal/cm²	cm
No	48	3	2	3	2.133	60.00
	48	6	2	3	4.267	84.85
	48	9	2	3	6.400	103.92
	48	12	2	3	8.533	120.00
	48	15	2	3	10.667	134.16
	48	18	2	3	12.800	146.97
	48	21	2	3	14.933	158.75
	48	24	2	3	17.067	169.71
	48	27	2	3	19.200	180.00
	48	30	2	3	21.333	189.74
Yes	48	3	0.1	3	0.107	13.42
	48	6	0.1	3	0.213	18.97
	48	9	0.1	3	0.320	23.24
	48	12	0.1	3	0.427	26.83
	48	15	0.1	3	0.533	30.00
	48	18	0.1	3	0.640	32.86
	48	21	0.1	3	0.747	35.50
	48	24	0.1	3	0.853	37.95
	48	27	0.1	3	0.960	40.25
	48	30	0.1	3	1.067	42.43

Appendix E – DC arc flash risk assessment

Table E.5 Arc flash incident energy and arc flash protection boundary distances for 120 V battery systems (Table F.3 in AS/NZS 5139)

Inter-string protection	DC arc voltage	Bolted fault current at point of activity	Arcing time	Multiplying factor	Incident energy	Arc flash boundary
	V	kA	s		cal/cm²	cm
No	120	3	2	3	5.333	94.87
	120	6	2	3	10.667	134.16
	120	9	2	3	16.000	164.32
	120	12	2	3	21.333	189.74
	120	15	2	3	26.667	212.13
	120	18	2	3	32.000	232.38
	120	21	2	3	37.333	251.00
	120	24	2	3	42.667	268.33
	120	27	2	3	48.000	284.60
	120	30	2	3	53.333	300.00
Yes	120	3	0.1	3	0.267	21.21
	120	6	0.1	3	0.533	30.00
	120	9	0.1	3	0.800	36.74
	120	12	0.1	3	1.067	42.43
	120	15	0.1	3	1.333	47.43
	120	18	0.1	3	1.600	51.96
	120	21	0.1	3	1.867	56.12
	120	24	0.1	3	2.133	60.00
	120	27	0.1	3	2.400	63.64
	120	30	0.1	3	2.667	67.08

Appendix E – DC arc flash risk assessment

Table E.6   Arc flash incident energy and arc flash protection boundary distances for 350 V battery systems (Table F.4 in AS/NZS 5139)

Inter-string protection	DC arc voltage	Bolted fault current at point of activity	Arcing time	Multiplying factor	Incident energy	Arc flash boundary
	V	kA	s		cal/cm²	cm
No	350	3	2	3	15.556	162.02
	350	6	2	3	31.111	229.13
	350	9	2	3	46.667	280.62
	350	12	2	3	62.222	324.04
	350	15	2	3	77.778	362.28
	350	18	2	3	93.333	396.86
	350	21	2	3	108.889	428.66
	350	24	2	3	124.444	458.26
	350	27	2	3	140.000	486.06
	350	30	2	3	155.556	512.35
Yes	350	3	0.1	3	0.778	36.23
	350	6	0.1	3	1.556	51.23
	350	9	0.1	3	2.333	62.75
	350	12	0.1	3	3.111	72.46
	350	15	0.1	3	3.889	81.01
	350	18	0.1	3	4.667	88.74
	350	21	0.1	3	5.444	95.85
	350	24	0.1	3	6.222	102.47
	350	27	0.1	3	7.000	108.69
	350	30	0.1	3	7.778	114.56

Appendix F

Frequently asked questions?

F1   Is it possible to omit a means of earthing for a socket-outlet on an EESS which continues to provide power in island mode or as a UPS supply?

The conditions for the provision and safe use of such a socket-outlet would lie with the manufacturer's design.

It is not recommended that such a socket-outlet is provided with a means of connection to an electrical installation or a permanently installed item of equipment. The reasons for this are:

(a) from the manufacturer's perspective:
  (i) BS EN IEC 62040-1 has certain conditions for the omission of protective earthing to the output circuit when the supply is removed.
  (ii) if a socket-outlet is to operate as IT (isolated from earth) in the emergency power mode, this may not be suitable for Class I equipment, , particularly where this is interconnected with other equipment or external systems such as those for telecommunications, antennae or television distribution systems. Such arrangements may present a shock hazard if not supplied by an earthed system.
(b) from the installer's perspective:
  (i) designers and installers select equipment complying with relevant standards, or having equivalent safety, meeting the requirements of BS 7671.
  (ii) an island mode or UPS-configured EESS is a generator acting as a switched alternative to the grid supply. Where the EESS is used as a switched alternative to the grid supply to part or all of a fixed electrical installation, it should comply with the earthing and bonding requirements of Section 551 of BS 7671, and cannot rely on the distributor's means of earthing.
  (iii) a BS 1363-2 socket-outlet may be used to connect multiple items of equipment. Separated systems in installations not under the control or supervision of skilled or instructed persons should not supply more than one item of equipment (see Regulation 413.1.3 of BS 7671).

F2   How does a local load automatically consume all of the generated current from a local grid-connected inverter, and the balance from the grid?

Voltage, phase and frequency are synchronized by the inverter, to connect with the grid. To control the power flow to the load from the grid, the inverter varies these parameters. The inverter monitors the grid voltage and adjusts its output voltage to a slightly higher value, so that the power drawn from the inverter is controlled. This increase in voltage is controlled based on the grid voltage and power output of the system on the DC side of the inverter.

Current flows from the grid into the load, and in doing so, the grid voltage will drop slightly according to the source impedance (Z_e) of the grid. In most cases, it will be a fraction of an ohm; but it will not be zero. When an inverter is connected, it will try to drive current into the grid, slightly raising the grid voltage, according to how much current it attempts to drive. The power being consumed in the installation is split between the grid and the inverter, because of the source output impedance. If the source impedance were zero, the split of power between the inverter and the grid would become indeterminate.

F3 With multiple parallel energy sources with grid or island mode – for example, PV, wind or battery storage – how is the current delivery co-ordinated? Are they all synchronized to deliver their combined total, or co-ordinated so as not to exceed a specific value?

For micro-generators (up to and including 16 A per phase) in parallel with public LV distribution networks, G98 states that:

> the total aggregate capacity of the Micro-generators (both non-Electricity Storage and Electricity Storage) is less than or equal to 16 A per phase

This means, to G98, that multiple generators shall not exceed a total of 16 A.

G99 requirements for the connection of generation equipment in parallel with public distribution networks state:

> The maximum aggregate capacity of Power Generating Modules that can be connected to a single phase supply is 17 kW. Power Generating Facilities with a capacity above 17 kW per phase are expected to comprise three phase units.

 **Appendix G**

General example schematics

This Appendix contains example schematics for practical EESS, as follows:

(a) Figure G.1 Example of EESS added to simple solar PV installation using DC coupling, configured for connected mode operation only

(b) Figure G.2 Example of EESS with local generation, AC-coupled, configured for connected mode operation only, with separate EESS inverter and battery unit

(c) Figure G.3 Example of EESS with local generation, AC-coupled, configured for both connected mode and island mode operation

(d) Figure G.4 Example of EESS with multiple sources of local generation, AC-coupled, configured for both connected mode and island mode operation

(e) Figure G.5 Example of EESS with local renewable generation and diesel generator, intended for grid-independent operation

(f) Figure G.6 Example integration of a vehicle to grid system and an EESS (TN system), configured for both connected mode and island mode operation, and for the vehicle to supply power in island mode

(g) Figure G.7 Example integration of a vehicle to grid system, an EESS (TN system) and local generation, configured for both connected mode and island mode operation, and for the vehicle to supply power in island mode

(h) Figure G.8 Example integration of a vehicle to grid system and an EESS (TT system), configured for both connected mode and island mode operation, and for the vehicle to supply power in island mode

Note: These are intended to give context to the Code of Practice, rather than provide specific wiring diagrams for real installations.

Appendix G – General example schematics

Figure G.1 Example of EESS added to simple solar PV installation using DC coupling, configured for connected mode operation only

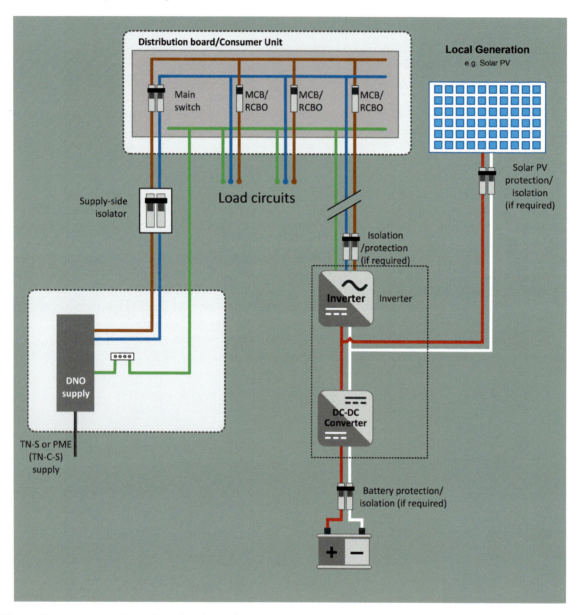

Note: This system must be configured so that it does not operate in island mode.

Appendix G – General example schematics

Figure G.2 Example of EESS with local generation, AC-coupled, configured for connected mode operation only, with separate EESS inverter and battery unit

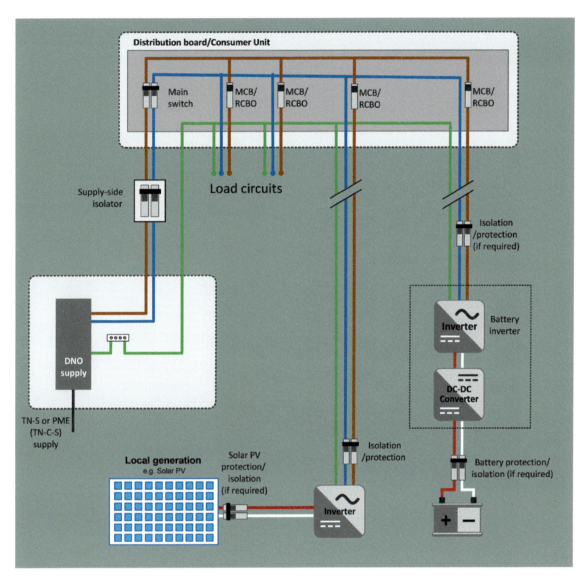

Note: This system must be configured so that it does not operate in island mode.

Appendix G – General example schematics

Figure G.3 Example of EESS with local generation, AC-coupled, configured for both connected mode and island mode operation

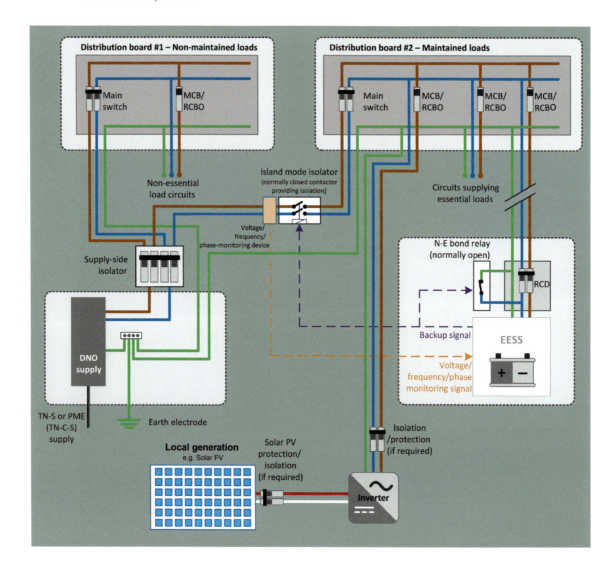

Appendix G – General example schematics

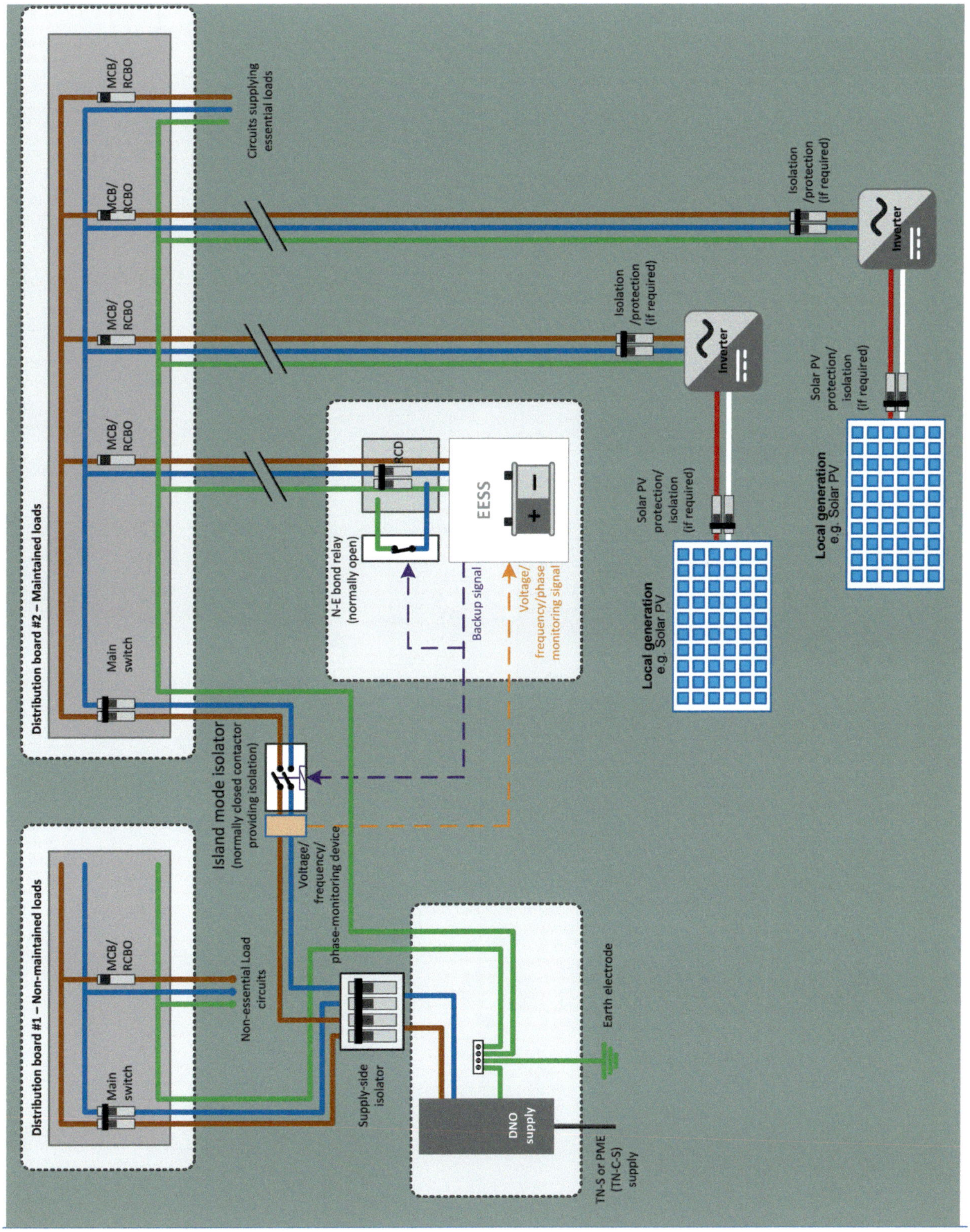

Figure G.4 Example of EESS with multiple sources of local generation, AC-coupled, configured for both connected mode and island mode operation

Appendix G – General example schematics

Figure G.5 Example of EESS with local renewable generation and diesel generator, intended for grid-independent operation

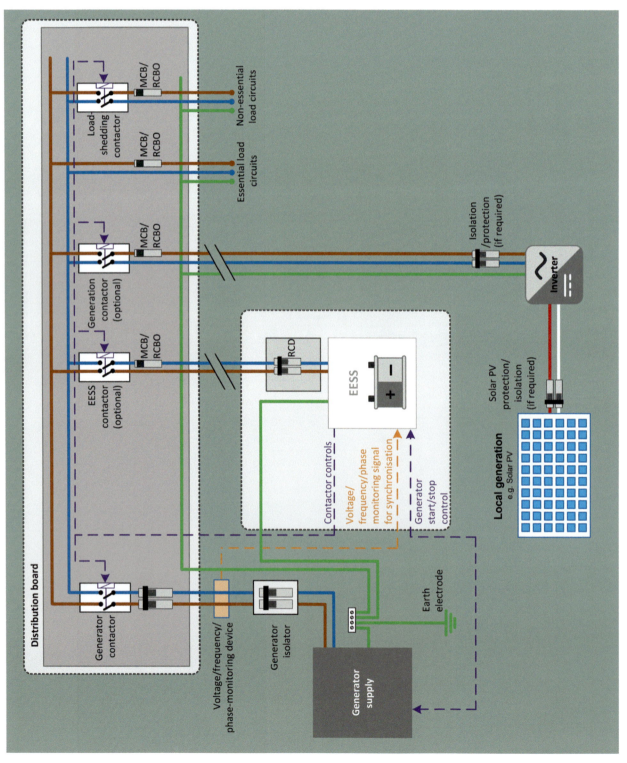

Note: Where the N-E bond is provided in the generator, the system would be configured to shut down the inverters when the generator is isolated or disconnected. Alternatively, the system may either be designed to operate as an IT system when the generator is isolated or disconnected, or the N-E bond made at the distribution board.

Appendix G – General example schematics

Figure G.6 Example integration of a vehicle to grid system and an EESS (TN system), configured for both connected mode and island mode operation, and for the vehicle to supply power in island mode

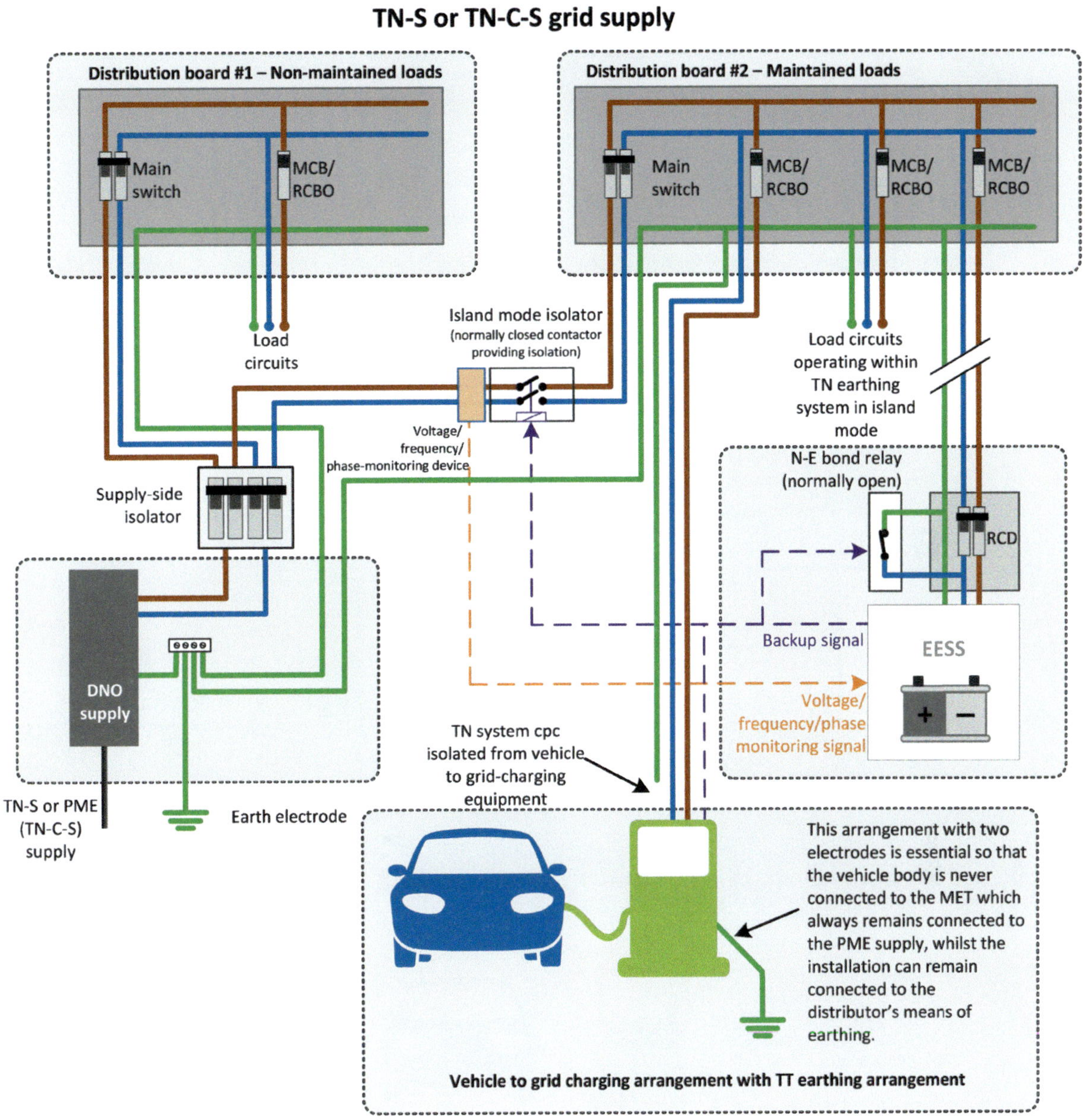

TN-S or TN-C-S grid supply

Note: Installation, commissioning and notification of this arrangement will follow a G99 route if the total generation (EESS + vehicle to grid) exceeds 16 A per phase.

Appendix G – General example schematics

Figure G.7   Example integration of a vehicle to grid system, an EESS (TN system) and local generation, configured for both connected mode and island mode operation, and for the vehicle to supply power in island mode

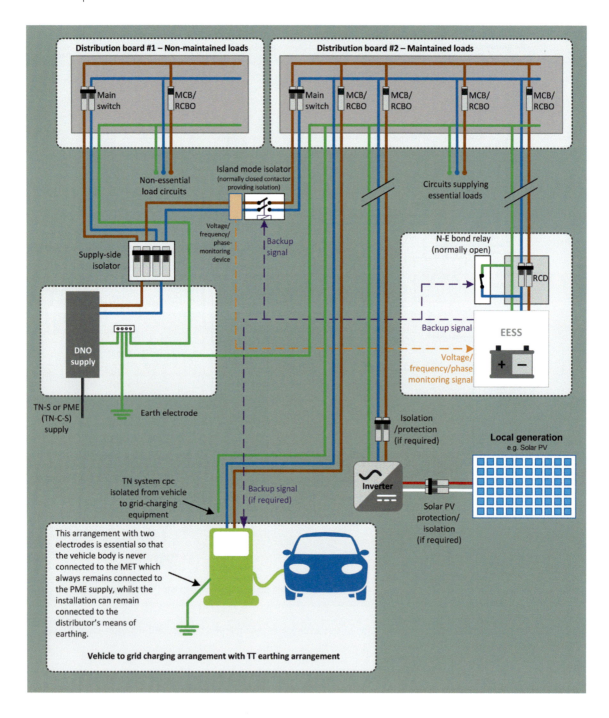

Appendix G – General example schematics

Figure G.8 Example integration of a vehicle to grid system and an EESS (TT system), configured for both connected mode and island mode operation, and for the vehicle to supply power in island mode

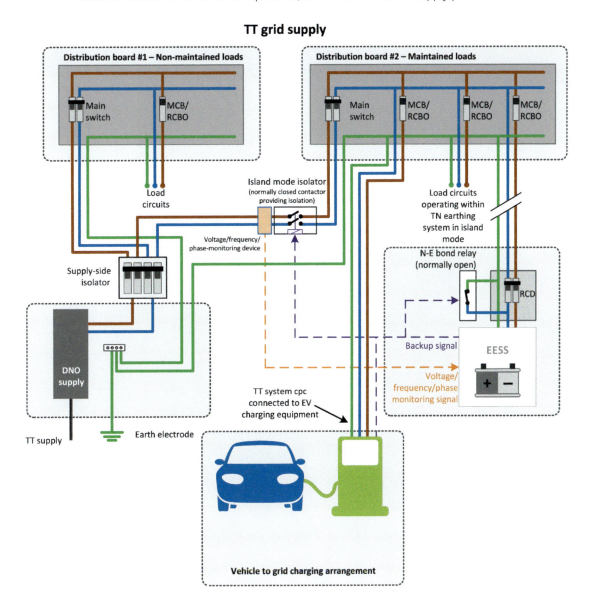

Index

Index

Index

Index

Index

Index

Complement your learning from our Codes of Practice with online courses from the IET Academy

The IET Academy provides engaging, interactive e-learning, covering a range of courses for all stages of your career to help you stand out from the engineering crowd.

E-learning gives you the opportunity to study at your own pace, in your own time. Flexible mobile access allows you to take a course in full or in shorter bite-size units, and course content is interactive and in-depth, providing the same level of learning you'd get from in a classroom.

Our courses are designed specifically for engineers, providing learning that is relevant to what you do. Courses are available in the following areas:

| ELECTRICAL | COMMUNICATIONS | POWER | PROFESSIONAL SKILLS | SAFETY & SECURITY | TRANSPORT | WIRING REGULATIONS |

Find out more at **theiet.org/academy-mp**

The Institution of Engineering and Technology (IET) is registered as a Charity in England and Wales (No. 211014) and Scotland (No. SC038698).
The Institution of Engineering and Technology, Michael Faraday House, Six Hills Way, Stevenage, Hertfordshire SG1 2AY, United Kingdom.